The Growth of
a Refining Region

INDUSTRIAL DEVELOPMENT AND
THE SOCIAL FABRIC Volume 4

Editor: Glenn Porter, Director, *Regional Economic History Research Center, Eleutherian Mills—Hagley Foundation, Greenville, Delaware*

INDUSTRIAL DEVELOPMENT AND THE SOCIAL FABRIC

An International Series of Historical Monographs

Series Editor: Glenn Porter
Director, Regional Economic History Research Center,
Eleutherian Mills—Hagley Foundation, Greenville, Delaware

Volume 1. FORUMS OF ORDER: The Federal Courts and Business in American History
Tony Allan Freyer, *University of Arkansas, Little Rock*

Volume 2. COMPETITION AND REGULATION: The Development of Oligopoly in the Meat Packing Industry
Mary Yeager, *University of California, Los Angeles*

Volume 3. KEEPING THE CORPORATE IMAGE: Public Relations and Business, 1900–1950
Richard S. Tedlow, *Graduate School of Business Administration, Harvard University*

Volume 4. THE GROWTH OF A REFINING REGION
Joseph A. Pratt, *School of Business Administration, University of California, Berkeley*

Volume 5. BUSINESS AND GOVERNMENT IN THE OIL INDUSTRY: A Case Study of Sun Oil, 1876–1945
August W. Giebelhaus, *Georgia Institute of Technology*

Volume 6. PIONEERING A MODERN SMALL BUSINESS: Wakefield Seafoods and the Alaskan Frontier
Mansel G. Blackford, *Ohio State University*

To my parents, who lived through this history,

and my wife, who lived through the writing of it.

The Growth of a Refining Region

by JOSEPH A. PRATT
School of Business Administration
University of California, Berkeley

Greenwich, Connecticut

Library of Congress Cataloging in Publication Data

LMS Pratt, Joseph A
The growth of a refining region.

(Industrial development and the social fabric;
v. 4)
Bibliography: p.
Includes index.
1. Petroleum industry and trade—Gulf
coast (United States) 2. Petroleum—Gulf
coast (United States)—Refining. I. Title.
II. Series.
HD9567.A13P7 338.4'766553'0976414 77-7797
ISBN 0-89232-090-7

165 West Putnam Avenue
Greenwich, Connecticut 06830

ISBN NUMBER: 0-89232-090-7
Library of Congress Catalog Card Number: 77-7797
Manufactured in the United States of America

CONTENTS

Tables

Maps

Acknowledgments

This book began too many years ago as a dissertation in the history department at Johns Hopkins under the direction of Louis Galambos. His guidance left a permanent mark on both the manuscript and its author. The "Galambos seminar" played an invaluable role in the planning, the writing, and the revision of the dissertation. To its members, past and present, my thanks. Another Hopkins professor, Willie Lee Rose, commented on the final draft with clarity mixed with kindness.

In the three years since leaving Hopkins, my efforts to reorganize, rewrite, and revise this manuscript have been aided by the editor of this series, Glenn Porter, who is a fellow native of the Houston area. My colleagues in the Political, Social and Legal Environment of Business field at the School of Business Administration of the University of California, Berkeley, have helped me to view history in a broader context. John O. King of the University of Houston was most helpful. At the beginning of this project, he guided me to crucial sources; at the end, he read the final draft.

The most important of these sources was the Texaco Archives. Kenneth McCullam, the historian and archivist at Texaco, was a model of cooperation and good will. Without his aid, this book would have taken a very different form. Without the benefit of his careful reading of the final draft, several mistakes would have been overlooked. Thomas Nanney and Lois Schuermann at the American Petroleum Institute were equally helpful. Numerous archivists provided assistance. Covington Rodgers at the University of Houston deserves special thanks, as does Beth Ray at the Woodson Research Center at Rice University.

I have been blessed with excellent typist-editors. My wife, Suzy, worked overtime on the early drafts of this manuscript. The final drafts were completed by Virginia Douglas, Betty Kendall, Ellen McGibbon, and Helen Way at the Institute of Business and Economic Research (IBER) at the University of California, Berkeley.

Support for my research was provided by IBER, by a Regents Junior Faculty Summer Fellowship from the University of California, Berkeley, by the School of Business Administration at Berkeley, by a Resources for The Future Research Fellowship in Natural Resources, by a Rovensky Fellowship in Business History from the Lincoln Educational Foundation, and by a working wife. The traveling required to get to a variety of sources was made much easier by a number of people with guest rooms or comfortable couches. These included various Pratts, Hardins, and Murphys, and the Sosnas, Longs, Prevatts, Moores, Galamboses, Porters, Van Courts, and Wilcoxes.

Several acknowledgments remain in addition to those in the dedication. Allan Matusow introduced me to history; Henrietta Larson made me think of myself as a historian; and Paul Blair, undoubtedly the best in the field, helped me make it through the dog days of several Augusts.

Part I

Growth, the Region, and Refining

Introduction

Throughout the twentieth century, the Gulf Coast of the United States has been one of the largest centers of petroleum refining in the world. The construction of refineries in this area began immediately after the discovery of the spectacular Spindletop oil field near Beaumont, Texas in January 1901. The Gulf Oil Corporation and Texaco, two companies that originated at Spindletop, built the first large plants at nearby Port Arthur, Texas.[1] With the subsequent opening of numerous other southwestern oil fields, more petroleum companies migrated to the Gulf Coast, and the refining complex gradually spread southward to Corpus Christi and eastward to New Orleans. By the 1970s this area contained almost 35 percent of the nation's refining capacity.[2] My account of its growth focuses on the historical and geographical center of Gulf Coast refining, the approximately 100 miles of coastline from Houston to Port Arthur. This region, which I will refer to throughout this study as the upper Texas Gulf Coast,[3] contains both the greatest concentration of refineries and Houston, the largest city on the Gulf Coast. My primary objective is to analyze some of the most important ties between the expanding refining complex and the developing region.

Even before the discovery of oil, the rise of Houston as a rail center and the expansion of timber and cotton production in surrounding areas had created a thriving economy. Spindletop introduced a new era of development, however, as the scramble for oil transformed the region. The production and transportation of oil quickly became major industries. More gradually, petroleum refining gained in importance. Before World War I the Port Arthur area contained all of the largest regional refineries, but this began to change after 1914 with the opening of the Houston Ship Channel, which attracted numerous plants to the Houston area. The next three decades witnessed the construction of new refineries throughout the upper Texas Gulf Coast, and in the 1920s and 1930s refining played a more prominent role in the regional economy than at any other time. These plants remained cornerstones of the regional economy after World War II, but the rise of petrochemical production and numerous other industries brought a decline in their relative importance. Even as the regional economy matured and diversified, however, the effects of the historical dominance of refining remained, for the most important of the new industries that arose in the postwar era had strong ties to the oil industry.

The imprint of petroleum on the pattern of regional development thus stands out vividly, like a brightly lighted refinery against the early morning sky on the coastal plain. The oil industry's expansion was the prime determinant of the pace of the region's twentieth century development. In addition, changes in the structure and the technology of the industry greatly affected the course of industrialization. Although all phases of the industry were important, refining left its mark most prominently on the region. As the fixed hubs of the worldwide operations of vertically integrated companies, the large Gulf Coast refineries provided permanent and steadily expanding linkages between these expansive companies and the region. The historical magnitude of these ties justifies the label "refining region," and their study explains much about the industrial development of the upper Texas Gulf Coast.

Through these linkages flowed a potent impetus for regional growth. As a primary supplier of an ever increasing demand for energy, the petroleum industry expanded steadily for most of the century. The regional refineries grew apace, producing a variety of goods that were essential to the modern economy. Gasoline and lubricants for the operation of internal combustion engines, asphalt for an expanding highway system, and the petroleum derivatives used in synthetic rubber for tires all tied the fate of the refineries to that of an automobile-related group of industries that has been one of the most dynamic sectors of the national economy for most of the twentieth century. Other refined products—from heating oils to the raw materials used in the production of petrochemicals—were also much in demand. The Gulf Coast refining region expanded more rapidly than other refining centers because it provided a substantial share of the refined goods demanded by the urban-industrial centers in the Northeast as well as by the industrializing Southwest. Refining was thus an impressive engine of growth that helped power more than seventy-five years of regional development at a pace that was unusual even by the standards of a nation accustomed to rapid economic growth.

Houston was the focus of much of this expansion, and its swift rise among the hierarchy of the nation's largest cities provides a fitting symbol for the development of the coastal region. In 1900, Houston's population of less than 45,000 made it the eighty-fifth largest city in the United States. By the mid-1970s, its approximately 1.5 million people made it the fifth most populous city in the nation. Throughout this period it has become increasingly important as a center for energy-related activities. Indeed, Houston's emergence in recent years as the unofficial "energy capital" of the nation is the logical culmination of over three-quarters of a century of oil-led development.

The steady and often spectacular expansion of Houston has been only the most visible result of a broader change in the distribution of population and manufacturing in the coastal region. Not just Houston, but numerous other smaller cities along the upper Texas Gulf Coast, prospered as refining and other oil-related activities helped attract several generations of farmers and workers from the interior. The result was a marked shift of people and industrial activity from the interior to the coast, a shift that both accompanied and made possible the rapid urbanization and industrialization of the refining region.

The works of several scholars concerned with these two closely related processes suggest theoretical insights into the evolution of the economy of the upper Texas Gulf Coast.[4] The discovery of oil and the rapid construction of numerous large refineries represented a sudden, dramatic shift in the course of the development of the coastal region. "Growth pole" theory, which is most commonly applied to developmental problems in modern regions that have been bypassed by growth, deals specifically with the implications of this type of thrust toward a new level of economic activity.[5] This theory suggests that growth can be induced in less developed regions by the introduction of "propulsive industries," which are characterized as relatively new, large, rapidly growing, innovative, technologically advanced industries with strong linkages to other segments of the economy. Such linkages will, in theory, foster an industrial complex, which, when successful, will encourage development in the remainder of the regional economy.[6] The historical experience of the upper Texas Gulf Coast closely mirrors the process described by growth pole theory, with one very important qualification. This region had not been bypassed by growth before 1901, for decades of previous development had prepared it well to respond to the "propulsion" generated by the discovery of oil.

Although growth pole concepts help to understand the impact of the initial intrusion of oil on the Gulf Coast, a second related body of theory suggests the broader outlines of the long-run development of the region. Wilbur Thompson has suggested the following stages of urban growth: (1) the expansion of an export-based economy with reliance on a single export; (2) the rise of competing exports to form an export complex; (3) economic maturation after the substitution of imports with regionally manufactured goods; (4) the growth of a regional metropolis as an exporter of services.[7] As with most such stage theories, this model represents as a simple progression a process that is, in reality, not necessarily continuous or clearly divided. Nonetheless, these stages provide a useful overview of the twentieth century development of the refining region and its "metropolis," Houston.

Against the background provided by such theories, Part I of this book presents a century-long overview of the course of development in the refining region. Chapter I, "The Region Before Refining: 1870–1901," contains a summary of economic growth in the thirty years before the discovery of oil. The second chapter, "The Creation of a Refining Region: 1901–1916," analyzes the economic impact of the discovery of oil and of the initial development of the oil industry. Chapter III, "The Heyday of Refining: 1916–1941," describes a period when oil refining was the region's primary manufacturing industry, accounting for most of its exports, up to half of its value added by manufacturing, and only a slightly smaller fraction of all employment in manufacturing. In its heyday, refining provided the primary economic propulsion for the development of the coastal region while also encouraging the expansion of other related industries. During and after World War II, expansion of the resulting oil-related complex of industries led the way toward a more diversified economy, and the outlines of this postwar development are traced in Chapter IV, "Refining in a Maturing Regional Economy: 1941–1960s." This chapter then concludes Part I with a summary of the broad changes in the location of population and manufacturing that accompanied a century of regional growth.

Whereas Part I is most concerned with the economic and geographical changes brought about by regional development, Part II focuses on important components of the institutional framework within which development occurred. The last four chapters examine the evolution of those institutions that exerted the greatest control over the growth process: vertically integrated oil companies and other large corporations in the oil-linked complex of industries (Chapter V); refinery-based workers' organizations (Chapter VI); and a variety of government agencies

(Chapters VII and VIII). Taken as a whole, Part II depicts historical changes in the division of labor among these institutions and in the relative power of each.

The institutions most influential in shaping development were the large, vertically integrated oil companies that controlled much of the regional and national oil industry. Gulf Oil, Texaco, Mobil Oil, Exxon, Shell Oil, and other similar companies brought to the region a new scale of economic organization. As they expanded, those parts of the regional economy most directly tied to them evolved into distinctive, oil-related sectors that operated differently than much of the remainder of the economy.

Economist Robert Averitt has discussed the position of such large firms in the post-World War II era in *The Dual Economy.* Averitt makes the following distinctions between the center firms and the peripheral firms that together form the dual economy:

> The center firm is large in economic size as measured by number of employees, total assets, and yearly sales. It tends toward vertical integration (through ownership or informal control), geographic dispersion (national and international), product diversification, and managerial decentralization. Center firms excel in managerial and technical talent; their financial resources are abundant. Their cash flows are large, particularly during prosperity; their credit ratings are excellent. Center managements combine a long-run with a short-run perspective. Short-run considerations are entertained at the lower levels of the managerial hierarchy, while long-run planning is the perquisite of top management. Their markets are commonly concentrated. Taken together, center firms make up the center economy.
>
> The periphery firm is relatively small. It is not integrated vertically, and it may be an economic satellite of a center firm or cluster of center firms. Periphery firms are less geographically dispersed, both nationally and internationally. Typically, they produce only a small line of related products. Their management is centralized, often revolving around a single individual.[8]

The concept of dualism encompasses much of the distinctive nature of modern American development, and it is thus a useful starting point for the analysis of the economic, social, political, and environmental impact of rapid growth in a regional economy increasingly dominated by an oil-related complex of industries composed of numerous "center firms."

Within the region this oil-related sector could usually outbid other industries for labor and services. But the crucial difference for the oil industry was its early and sustained access to national and international markets. It sold its regionally produced goods throughout the world, and it could tap sources of capital and managerial and technical talent that were beyond the reach of most other regional firms. Texaco and

Gulf, for example, quickly outgrew their regional origins to become nationally active firms. Even in their earliest years, however, Gulf and Texaco did not rely strictly on the local economy for financial services, for refinery supplies, for construction, or even for most of their skilled labor. Instead, they acquired goods and services primarily from a national market made up of a large, specialized group of petroleum-related industries that had originated in the early eastern oil fields and spread throughout the nation with the discovery of other oil fields and the construction of additional refineries. Similarly, the products of the large refineries along the Gulf Coast found their primary markets in the urban centers of the East Coast, not within the region itself. The large refineries were located in the coastal region, but in important respects, they were not of the region. The national petroleum market provided many goods and services, eastern sources of investment controlled the expansion of the firms, and the sophisticated technological advances that revolutionized the refineries were borrowed from outside the region. Control of most of the factors that influenced the growth of the region's major industry, refining, thus generally lay in the distant headquarters of nationally and internationally active corporations.

As the oil industry expanded, it encouraged similar shifts in the scale and structure of other important "regional" institutions. Perhaps the most significant such shift was the organization of workers in the giant coastal refineries into strong, independent, nationally active, industrial labor unions. After gaining much influence over the conditions of labor in the refineries, these organizations altered the distribution of both economic and political power in the region. Certain political institutions also became more complex and more powerful as they responded to both the needs of the oil industry and the problems created by its expansion. For labor and political institutions, as for the large oil companies, one significant consequence of these changes was that national and international institutions exerted increasing control over a wide range of decisions that greatly affected the regional political economy.

Control over the growth process thus became more centralized and beyond regional influence. The decision to expand a refinery took place in New York or Pittsburgh. The determination of regional wage rates was influenced by negotiations between union headquarters in Denver and national officials of the oil companies. From Washington came decisions affecting a wide range of refinery-related matters, from the size of the owner company to the volume and type of effluent discharges allowed. In spite of this shift in the locus of decision making, however, many of the consequences of the sustained growth of refining were felt primarily within the area immediately surrounding the refineries. Thus, as the upper Texas Gulf Coast became a "refining region"—that is, as it gained a functionally defined identity within the national petroleum industry and the national economy—it paradoxically lost much of its ability to affect the nature and the pace of its own evolution.

The steady shift in the locus of economic and political decision making to state and national institutions raises a difficult analytical problem: How should a "regional" study deal with the growing importance of these external influences on the region's development? In this book, I have taken an institutional rather than a geographical approach to this question. That is, I have generally attempted to follow the shifts upward in power to institutions outside of the region, rather than to discuss the implications of these shifts for regional institutions. As a result, this history of the Gulf Coast refining region devotes considerable space to discussions of the histories of the large oil companies; the Oil, Chemical, and Atomic Workers Union; the Texas Railroad Commission; and even the United States Department of State. It is thus regional history very broadly defined.

The first step, nonetheless, was to delineate what constituted the core region in question. For my purposes, the defining characteristic of the upper Texas Gulf Coast has been its historical role as an important location for large petroleum refineries serving national markets. I have thus defined the boundaries of this region so as to include the two largest centers of refining, the Port Arthur area and the Houston area. Of course, the impact of the growth of refining was not neatly isolated in these two areas. Although focused there, the economic effects of refinery expansion were felt in much of the remainder of the Texas-Louisiana Gulf Coast, as well as throughout the national economy. The refineries also exerted a powerful influence on the labor market throughout a broad section of eastern Texas and western Louisiana. The political influence of their owners was felt primarily outside the region proper, at the state and national levels. Finally, the environmental consequences of refinery growth directly affected the air and water systems that extended both inland from the coastal region and out into the Gulf of Mexico. Each problem discussed thus involves a slightly different "region," although all include the core refining area on the upper Texas Gulf Coast. All of these regions are useful in analyzing various aspects of growth. They are not, however, sufficiently broad, since all were influenced by larger systems of which they were a part. National and international economic, political, and environmental systems greatly influenced regional development, and I have tried to discuss changes in the core region in the context of these broader systems.

Whenever possible, I have also tried to provide a comparative perspective with the inclusion of information about other regions. In contrast to the older industrial centers in the Northeast, the refining region industrialized primarily in the twentieth century, and its development was thus relatively unimpeded by locational arrangements previously established in response to older and very different economic patterns and transportation systems. The region's evolution most closely resembled that of Los Angeles,[10] but it also was similar in important ways to the Detroit area, where the automobile industry shaped much of the modern growth process.[11] The region around Pittsburgh, with its strong historical ties to the steel industry, offers still another useful counterpoint for the study of the refining region.[12] As with Pittsburgh and Detroit, a distinctive characteristic of the Houston area was the extent to which a single industry shaped development.

Despite such broad similarities, there were, however, significant variations attributable to the differences between refining and other manufacturing industries. Compared to most other industries, refining was an excellent inducer of growth because it encouraged the rise of the natural gas and petrochemical industries and the expansion of numerous supply firms. The refineries also spawned a distinctive labor market, since they were among the most capital-intensive and technologically advanced of all modern manufacturing concerns. The refineries probably differed little from other large manufacturing plants in influencing local politics. But the large oil companies were very active at the state and national levels. The impact of the refineries on the regional environment was also somewhat different from that of other manufacturing concerns because the processing and transporting of oil gave rise to especially difficult water and air pollution problems. On balance the similarities in the process and consequences of rapid, sustained growth in all of these regions were more important than the differences, but it should be recognized that the petroleum industry in general and refining in particular lent a distinctive character to the Houston area.

My book seeks to capture a part of this distinctive character. It is not a comprehensive economic history of the coastal region.[13] Nor is it a history of Houston.[14] Although it stresses the crucial developmental role of several large oil companies, it is not an account of the internal evolution of these companies.[15] It is, above all, not an attempt to test existing theories of regional growth. Instead, it combines aspects of each of these approaches in an effort to sketch the broad outlines of both the economic and institutional impact of the growth of refining on the twentieth century development of the upper Texas Gulf Coast.

NOTES

1. The predecessor of Gulf Oil, the J. M. Guffey Petroleum Company, began operations at Beaumont in 1901. Texaco, which was created in April 1902, grew out of the Texas Fuel Oil Company, which originated soon after the Spindletop discovery.

2. American Petroleum Institute, "Number and Capacity of Refineries, by Refining District, as of January 1, 1970," *Petroleum Facts and Figures, 1971 edition* (Baltimore, 1971), p. 137.

3. I will use the phrase "upper Texas Gulf Coast" specifically to designate the focus of much of my analysis, the region from the Port Arthur area to the Houston area. For statistical purposes, I have defined the region to include Brazoria, Harris, Chambers, Galveston, Jefferson, and Orange counties in Texas and Calcasieu and Cameron parishes in Louisiana.

4. Theories of regional development are quite diverse—the complexity of the process involved allows theorists much leeway in extracting themes for study. For a discussion of various theories, see John Meyer, "Regional Economics: A Survey," *American Economic Review* LIII (March 1963), pp. 19-55. Also, Harry Richardson, *Regional Growth Theory* (New York, 1973). Douglass North, *The Economic Growth of the United States, 1790-1860* (Englewood Cliffs, N.J., 1961) used an export-based theory to analyze early nineteenth century growth. North and "classical theory" have been properly criticized for taking a narrow view of the growth process and ignoring important sectoral shifts within the economy. Export theory, of course, cannot explain all aspects of growth in all regions. It focuses, however, on a significant dynamic of change and suggests how this agent of change affects much of the remainder of the economy. This makes it especially useful for a study of a region in which the major economic activity has been the production of large quantities of refined goods for outside markets. See, also, Harvey Perloff, Edgar Dunn, Jr., Eric Lampard, and Richard Muth, *Regions, Resources, and Economic Growth* (Baltimore, 1960).

5. Growth pole theory embodies much that is common to traditional export-based theories of regional development. Brian J. L. Berry has summarized such an approach in analyzing what he labels "the classical theory of regional development." He cites five propositions at the heart of this theory: (1) a region is an interdependent part of a national economy; (2) regional growth is externally induced; (3) export sector growth translates into residentiary sector expansion; (4) growth takes place in a matrix of urban regions through which the economy is organized; (5) when growth is sustained over long periods of time, it results in the progressive integration of the space economy. See Brian J. L. Berry, *Growth Centers in the American Urban System,* vol. 1 (Cambridge, 1973), pp. 10-17; and Brian J. L. Berry and Frank Horton, *Geographic Perspectives on Urban Systems* (Englewood Cliffs, N.J., 1970), pp. 94-105. Good discussions of Francois Perroux's theory of growth poles and a short article by Perroux, entitled "Note on the Concept of Growth Poles," are found in David McKee, Robert Dean, and William Leahy (eds.), *Regional Economics: Theory and Practice* (New York, 1970). See, also, Niles Hansen (ed.), *Growth Centers in Regional Economic Development* (New York, 1972).

6. Tormod Hermansen, "Development Poles and Development Centers," in Antoni

Kuklinski (ed.), *Growth Poles and Regional Growth Centers* (Paris, 1972), pp. 14–51. This collection also contains several good essays on growth pole theory.

7. Wilbur Thompson, *A Preface to Urban Economics* (Baltimore, 1965), pp. 15–16.

8. Robert Averitt, *The Dual Economy* (New York, 1968), pp. 1–22.

9. I have discussed this problem in detail in "Regional Development in the Context of National Economic Growth," in Glenn Porter (ed.), *Regional Economic History: The Mid-Atlantic Area Since 1700* (Eleutherian Mills-Hagley Foundation: Greenville, Delaware, 1976), pp. 25–40.

10. Robert Fogelson, *The Fragmented Metropolis: Los Angeles, 1850–1930* (Cambridge, Mass., 1967).

11. Constantinos A. Doxiadis (ed.), *Emergence and Growth of an Urban Region: The Developing Urban Detroit Area,* 3 vols. (Detroit, 1966–67).

12. Pittsburgh Regional Planning Association, *Economic Study of the Pittsburgh Region,* 4 vols. (Pittsburgh, 1963); Roy Lubove, *Twentieth Century Pittsburgh* (New York, 1969).

13. A very thorough history of the region (as of 1955) is included in Joseph L. Clark, *The Texas Gulf Coast: Its History and Development,* 4 vols. (New York, 1955).

14. For such a history, see David McComb, *Houston: The Bayou City* (Austin, 1969).

15. An excellent history of one of the major oil companies active in the region is Henrietta Larson and Kenneth Porter, *History of Humble Oil and Refining Company: A Study in Industrial Growth* (New York, 1959).

Chapter I

The Region Before Refining: 1870–1901

January 1901 marked the beginning of oil-induced growth on the upper Texas Gulf Coast. But before the discovery of oil at Spindletop and the subsequent construction of large refineries nearby, the coastal economy was healthy and dynamic. During the last half of the nineteenth century, the regional railroad system expanded steadily, and in the 1870s and 1880s these local railroads were tied into transcontinental systems. Improved transportation encouraged the production of timber, rice, and cotton, and these activities spawned distinctive patterns of development in the coastal economy.

The growth of the oil industry after 1901 did not eliminate these patterns. It merely submerged them beneath the distinctive economic system required for the production, transportation, and refining of petroleum. Oil brought a new scale, pace, and direction, as well as a measure of unity, to regional development, but it did not transform a collection of isolated farming villages into a modern industrial society. Pre-oil development was substantial, and it produced a pool of local industrial skills and capital that later could be transferred to the oil industry or employed in prosperous secondary industries. A description of the region's evolution before 1901 thus provides an essential background for understanding the subsequent impact of the growth of the oil industry.

A REGIONAL TRANSPORTATION REVOLUTION

The location and the resource endowment of the region shaped its early development. Like most of the Gulf Coast, the area from Houston to Port Arthur lies on a flat coastal plain which extends 30 to 50 miles up from the Gulf of Mexico (see Map 1.1). It is flat, wet, and hot, and sections of the area are marshlands. Settlers coming to antebellum Texas usually did not venture east of Houston on the coastal plain. Frederick Law Olmsted, who traveled through Texas in the 1850s, found little between Houston and New Orleans to encourage settlement: "Upon the whole, this is not the spot in which I should prefer to come to light, burn, and expire; in fact, if the nether regions . . . be a boggy country, the avernal entrance might, I should think, with good probabilities, be looked for in this region."[1] Neither the marshlands nor the sandy soils

Map 1.1. The Topography of the Coastal Area

on the coast attracted settlers, but people were drawn to the pine woods region to the north and to the blackland prairie to the north and west. The stands of pine trees and the fertile soils available for growing cotton were of value, however, only if lumber and cotton could reach markets. The coastal plain was a natural site for processing, trading, and shipping the products of the interior, but the lack of an adequate transportation system hampered the growth of manufacturing and commerce in the coastal region.

A system of rivers connects the coast with the interior, but none of the rivers was naturally well-suited for shipping. The Red River provided a useful waterway for parts of northern Texas. Cotton shipments could be sent down the Red River to the Mississippi and then to New Orleans, bypassing the upper Texas Gulf Coast. The extensive use of other rivers awaited improvements. When goods were shipped to the coast, problems resulted from the lack of adequate natural harbors. Even the harbor of the island city of Galveston was poor, and expensive dredging and jetties were necessary before it could handle large ships. Other port sites, including Houston, Port Arthur, and Beaumont, were inland and required improvement of waterways as well as harbors.

Despite such handicaps, the coastal region grew gradually in the first half of the nineteenth century. Olmsted described Houston as an energetic town of about 3,000; the area to the east was "but sparsely settled," principally by "herdsmen."[2] Further development was dependent upon better transportation and communication ties between the interior and the coast and between the coast and the rest of the nation.

Railroad promoters in the city of Houston took the lead in tying the interior's resources to the coast. In the 1850s the construction of six lines that reached out from Houston for distances up to ninety miles made that city the most important rail center in the state.[3] Although Houston had been an important transshipment point before the 1850s, its ability to attract railroads was crucial in establishing it as the region's dominant city. In 1860 it was the original starting point or terminus for 80 percent of the approximately 500 miles of track laid in Texas.[4] Its early rail connections foretold its subsequent nineteenth century development; the most important roads stretched west toward the cotton belt and south toward the Gulf of Mexico at Galveston.

The Civil War temporarily checked the further development of Houston and of the entire region. Although the war had little direct effect in Texas, it prevented for a time the next crucial step in transportation improvement, the linking of local lines into the national railway system.

The local railroads encouraged regional expansion centered around Houston, but they did not provide access to markets potentially available in other larger cities. As late as 1871 the entire state of Texas had only one interstate rail connection, a short road in East Texas that extended twenty miles across the border to Shreveport, Louisiana.[5]

The 1870s and 1880s brought a sudden and dramatic change in this situation. Rivalries between transcontinental lines helped expand the nation's rail system from almost 53,000 miles in 1870 to more than 192,000 miles in 1900, and Texas in general and Houston in particular benefited greatly from this expansion.[6] The driving force behind the surge of the 1870s was the intense rivalry among builders of transcontinental lines. Collis Huntington of the Southern Pacific Railroad, Jay Gould of the Texas and Pacific Railroad, and William B. Strong of the Atchison, Topeka and Santa Fe Railroad were the dominant figures in transcontinental expansion through Texas. Huntington's San Francisco to New Orleans road, completed by 1883, consolidated existing lines that traveled from west of San Antonio through Houston and on to New Orleans.[7] The Southern Pacific also had a north-south connection reaching up from Houston into the cotton belt. Gould's Texas and Pacific went through northern Texas en route to St. Louis. It did not greatly affect the coastal region, except that its Dallas-to-Shreveport-to-New Orleans line competed with the Southern Pacific for New Orleans traffic. The Atchison group built a line from Kansas south to Houston and Galveston with an important extension in central Texas. The transcontinental builders were not the only railroad men active in Texas during the period, and local development similar to that around Houston in the 1850s continued on a broader scale. But the transcontinental lines performed the essential function of providing connections between the local roads and the national transportation system. Goods could still be shipped from Texas to Shreveport in 1890, but by that time San Francisco, New Orleans, St. Louis, Chicago, and New York could also be reached by rail.

Building on its early dominance of Texas railroads, Houston strengthened its claim to rail supremacy in the coastal region during the period of transcontinental expansion. By 1901 it had rail connections with all parts of the state. In addition, the Southern Pacific made Houston an important station between San Francisco and New Orleans. While establishing strong new rail connections to the national economy, Houston retained and strengthened its ties to both the cotton belt in the blackland prairie and the timber region of East Texas. As an early historian of Southeast Texas concluded, "the foundation of Houston's permanent prosperity as a city was laid on the railroads."[8] On the eve of the Spindletop discovery, that foundation had been completed.

The counties to the east of Houston also benefited from the extension of the Southern Pacific's line to New Orleans. The completion of the Texas and New Orleans (T&NO) in the 1880s and its subsequent purchase by the Southern Pacific stimulated growth in the lumber mill towns of Beaumont, near the mouth of the Neches River, and Orange, near the mouth of the Sabine.[9] In 1882 the T&NO acquired another important local road, the Sabine and East Texas, which stretched more than 100 miles from the small Gulf port at Sabine Pass through Beaumont and on into the pine forests of East Texas.[10] The T&NO then extended this line to Dallas, where it connected with national systems. The Santa Fe system also came through Beaumont. In 1895 its Gulf and Interstate provided a link with Galveston. In addition, the Santa Fe built the Gulf, Interstate, and Santa Fe from above Beaumont into central Texas. This road—another connection to the pine forests—was financed by a Houston-based lumberman, John Kirby, and was later sold to the Santa Fe system.[11]

Kirby's experience repeated that of several earlier Houston promoters. He built the line primarily to transport his own lumber, and the road profitably fulfilled his business needs. But a system like the Santa Fe could offer him a good price for his railroad while tying it to a larger rail network. By selling out, Kirby, like the earlier owners of the T&NO, avoided the financial problems connected with running a short railroad in competition with a transcontinental system. At the same time, he still achieved his original goal of cheaper, more efficient transportation for his lumber company. In their haste to expand and to challenge rival systems, the transcontinentals sometimes paid inflated prices for existing roads.[12] In the hands of men like Kirby, these dollars were generally invested in the regional economy, encouraging further growth.

A more direct contribution to regional development came from the construction of the final railroad to enter the region before 1901, the Kansas City, Pittsburgh, and Gulf (later renamed the Kansas City Southern). Arthur Stilwell, a promoter of railroads around Kansas City, decided that rail shipment to the Gulf of Mexico followed by water shipment to the eastern United States would lower the transportation cost of the midwestern grain then being shipped overland by rail to the East. From 1889 to 1894 his road from Kansas City gradually made its way, through acquisition and construction, toward the Gulf. Stilwell originally proposed to purchase existing lines from Shreveport to Houston and on to Galveston, which he planned to make his southern terminus. But in 1895 he decided instead to create a new town, Port Arthur, as the Gulf port for the KCP&G.[13] He selected a site fifteen miles southeast of Beaumont and, with the backing of Dutch investors, began promoting the new town with advertisements in midwestern newspapers. Beginning in 1897, excursion trains from Kansas City, with connections to Chicago, brought trainloads of prospective settlers. Stilwell and his railroad company were determined to create a town. Advertising and promotional efforts attracted a small population, and the need to feed, clothe, and house the populace generated local commercial and agricultural activity. The laying out of the town, the completion of the railroad, and the building of a large grain elevator and a dock stimulated growth while preparing the city for the extensive trade in wheat, corn, oats, and timber which the owners hoped the KCP&G would attract. Whereas the railroads transformed Houston, helping to change a small trading city into a center of regional commerce, the KCP&G actually created Port Arthur, constructing the rudiments of a Gulf port on previously undeveloped marshland.

A port requires navigable access to the sea, and Port Arthur's prospects for growth depended on a further transportation improvement—the dredging of a ship canal across Lake Sabine to the Gulf. Promoters of a rival port site at nearby Sabine Pass used federal court orders to block temporarily the completion of the project. But Stilwell and his associates finally overcame these difficulties. The completion of a canal twenty-six feet deep across Lake Sabine in 1899 fulfilled Arthur Stilwell's dream of a Gulf port—Port Arthur—connected to Kansas City by rail.[14]

About seventy miles down the coast, the port of Galveston had been growing steadily for more than fifty years. Its railroad ties with Houston increased both its shipping volume and the demand for harbor improvements. Federal aid for the harbor came in 1874, and another federal grant financed the construction of jetties after 1886. Such improvements helped Galveston maintain its position as both the largest port and the largest city on the coastal plain of Texas.[15] But several factors undermined Galveston's future prospects. As an island city, it faced difficulties transporting goods to and from the mainland. It was also vulnerable to hurricanes. Houston's location forty miles inland eliminated both problems. As early as 1872, a federal grant for improvement of Buffalo Bayou, Houston's link to Galveston Bay and to the Gulf, marked the beginning of Houston's efforts to reach out to the sea. The Rivers and Harbors Act of 1899 approved funds for the completion of a twenty-five foot ship channel. Although Galveston retained its position as the largest Texas port in the last half of the nineteenth century,

Houston had thus begun to move down the path that would ultimately make it the major port on the Texas coast.[16]

The combination of railroad building and waterway improvements that marked the progress of both Houston and Port Arthur was an essential prerequisite for further regional growth. A rudimentary highway system connected the major cities in Texas and Louisiana in the late nineteenth century, but some of the roads were little more than cattle paths, and even the major highways were dirt roads of limited use for commerce. From 1870 to 1901, private promotional efforts by both railroads and land promoters and public aid in the form of land grants to railroads and congressional appropriations for rivers and harbors improvements built a transportation system that gave the growing coastal region much greater access to national and world markets.

REGIONAL POPULATION GROWTH BEFORE SPINDLETOP

The magnitude of the region's growth between 1870 and 1900 is suggested by the expansion of the population in the coastal region and in the interior areas connected both economically and geographically to the coast. Table 1.1 and Map 1.2 show a rapid increase in population throughout a large section of Texas and Louisiana. Since the ports on the coast served as outlets to the sea for most of this section's exports, their growth was closely connected with that of the interior. Table 1.2 suggests several trends in the population growth of these coastal cities and of the section as a whole. The cities were growing steadily, if not spectacularly, but they were not yet especially large in comparison to the surrounding regions. As of 1900, they accounted for 64 percent of the population of the coastal region, but only 11 percent of the sectional total. Within the coastal region, Houston's rapid growth from 1880 to 1900 was noteworthy, since it enabled that city to overtake its neighbor and rival, Galveston. But despite this expansion, as late as 1900, Houston had only 5 percent of the population of the entire section. At that time it was less than one-sixth the size of New Orleans, the major port and largest city on the Gulf Coast throughout most of the nineteenth century.

Further comparisons between Houston and other cities in the nation help place its growth and that of the coastal region in a national perspective. Cities that shared with Houston a population of between 40,000 and 50,000 in 1900 included Yonkers, New York; Norfolk, Virginia; Brockton and Holyoke, Massachusetts; Ft. Wayne, Indiana; Youngstown

Table 1.1. Sectional Population Shifts: 1870–1900

	1870	1880	1890	1900	% Change 1870-1900
Coast	50,500	77,900	104,900	161,900	221
Cotton	175,467	286,850	351,132	433,476	147
Mixed	53,259	89,088	107,135	140,856	164
Timber	72,800	94,700	121,900	189,000	160
Total	351,689	550,057	684,222	924,745	163
Total -Coast	301,198	472,157	579,322	762,845	153
U.S.	39,818,449	50,155,783	62,947,714	75,994,575	90

Coast - Brazoria, Harris, Chambers, Galveston, Jefferson, Orange counties in Texas. Calcasieu and Cameron parishes in Louisiana. Due to parish boundary changes during the period, the Calcasieu parish population was estimated.

Cotton - Freestone, Limestone, Falls, Leon, Robertson, Milam, Madison, Grimes, Brazos, Burleson, Lee Fayette, Washington, Waller, Austin, Colorado, Lavaca, Ft. Bend, Wharton, Jackson, Matagorda counties in Texas.

Mixed - (Timber and Cotton) - Cherokee, Anderson, Houston, Trinity, Walker, San Jacinto, Montgomery, and Liberty counties in Texas.

Timber - Shelby, Nacogdoches, Sabine, San Augustine, Angelina, Polk, Tyler, Hardin, Jasper, Newton counties in Texas. Beauregard, Jefferson Davis, Allen, Vernon, Sabine, De Soto parishes in Louisiana. Due to parish boundaries changes during the period, the Beauregard, Jefferson Davis, and Allen populations were estimated.

Note: The labels "coast," "cotton," "mixed," and "timber" are used to differentiate parts of the large section of eastern Texas and western Louisiana that surrounded the coast (see Map 1.2). These labels suggest the major economic activity of each of the regions in the interior as of the 1870s; the "mixed" region relied on both cotton and timber. Of course, in the late 19th century, many residents in all of the regions were subsistence farmers. Throughout *Part I,* I will refer back to these regional subdivisions within this section of Texas and Louisiana in an effort to show the broad shifts in sectional population encouraged by the growth of petroleum-related manufacturing in the coastal region.

Map 1.2. The Coastal Region with the Surrounding Regions

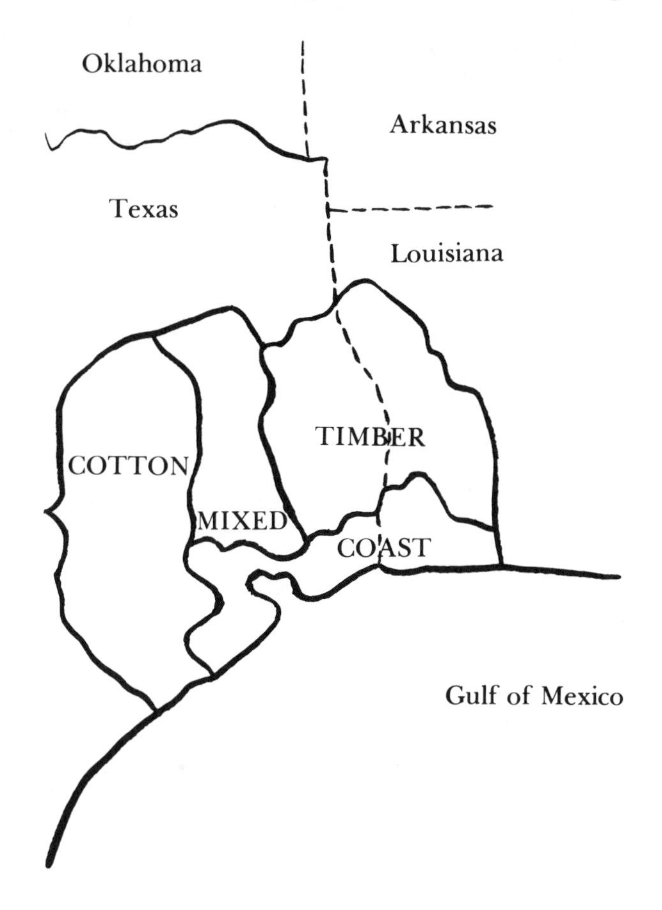

Note: The labels "cotton," "mixed," "timber," and "coast" will refer to these specific regions in the remainder of this study. "Section" will be used to designate the area encompassed by all four of these "regions."

and Akron, Ohio; Covington, Kentucky; Dallas, Texas; Saginaw, Michigan; Lancaster, Pennsylvania; Lincoln, Nebraska; and Waterbury, Connecticut. Among these cities, only Dallas matched Houston's growth in the twentieth century. Yet from 1870 to 1900, Yonkers, Youngstown, Akron, Dallas, Holyoke, Brockton, and Waterbury all grew at roughly the same rate as Houston.[17] In the three decades before the discovery of

Table 1.2. The Growth of Cities in the Coastal Region, 1870–1900

Cities in the Coastal Region:

	1870 Pop.	1900 Pop.	% Change 1870–1900
Houston	9,382	44,623	376
Galveston	13,818	37,789	173
Beaumont	-	9,427	-
Orange	-	3,835	-
Port Arthur	-	900	-
Lake Charles	-	6,680	-
TOTAL	23,200	103,254	345

Other Cities on the Texas-Louisiana Gulf Coast:

	1870 Pop.	1900 Pop.	% Change 1870–1900
New Orleans	191,418	287,104	50
Baton Rouge	6,498	11,269	73
Corpus Christi	2,140	4,703	120

Source: Bureau of the Census, Census of Population, 1870, 1880, 1890, 1900.

oil, Houston was a rapidly growing intermediate-sized city in a national urban hierarchy characterized by similar rapid growth at most levels.

Statistics on population alone neither establish the importance of a city within its section nor predict its subsequent growth. The historical course of population movement in the United States has been ever westward, and ranking by city size in 1870 and 1900 captures only a brief thirty-year part of this trend. But such a portrait does suggest an important consideration that is useful in understanding the pace of growth of Houston, of the Texas coastal region, and of the southwestern United States as a whole. Between 1870 and 1900 Houston was a prospering, dynamic city, but it was not yet even measured on the same scale as the established eastern and midwestern urban centers that had evolved gradually over a much longer period of time. Its rank as the eighty-fifth largest city in the nation reflected a general lack of development throughout the Southwest. From 1870 to 1900, the Texas coastal region, like the Southwest of which it was a part, grew rapidly. But as of 1900, its population was not yet comparable to that of the urban areas in the East and Midwest. At the beginning of the present century, slightly fewer people lived in the six largest cities in Texas (Houston, Galveston, San Antonio, Austin, Dallas, and Forth Worth) than lived in Newark, New Jersey.

DEVELOPMENT OF REGIONAL INDUSTRIES: 1870-1901

Similar conclusions emerge from an examination of the most important regional industries. The railroads themselves were the major developmental force, and Houston drew special economic benefits from their expansion. In 1891, more than 1,900 Houstonians worked in the construction and repair shops of the Southern Pacific and the Houston and Texas Central, which were among the leading industrial employers in the region.[18] According to the 1900 census, they accounted for 17 percent of Houston's salaried officials, 28 percent of its wage earners, and more than 30 percent of the total wages and salaries paid in manufacturing industries in 1900.[19] The railroads were also a major employer in Beaumont, where the division roundhouse on the Southern Pacific had between 300 and 450 workers in the late 1890s.[20]

The railroads paid higher wages than most other industries, but their impact on the regional labor force went far beyond wages. The skills required by the railroads often were transferable to a variety of other industries. The Texas and New Orleans alone employed 180 machinists, 259 other shopmen, 191 carpenters, 47 telegraph operators, and 154 office clerks in Texas in 1900.[21] The industrial and administrative capabilities of these and other railroad employees were important resources in a growing regional economy. The large railroads, which were the first modern corporations in the region, brought to the region a new scale of industrial operations that greatly augmented its supply of skilled laborers and managers.[22]

The railroads also brought new capital for investment. In addition to their personal resources, those who ran the major railroads often had strong financial and personal ties to other potential investors. Local enterprises could hope to attract financial backing from such men, even when they were connected to the region only through their financial ties with the area's railroads. The owners of these roads had an interest in encouraging such investments, since their long-run prosperity depended on the growth of sufficient local trade to support their businesses. On one level, the railroad's role in attracting regional investments consisted of such promotional efforts as the KCP&G's excursion trains and midwestern newspaper advertisements or the Southern Pacific's annual publication of *The Coast Country of Texas,* a promotional magazine that extolled the region's investment opportunities.[23] Probably of greater importance, however, were the more personal "promotional campaigns" which consisted of prominent railroad men investing their own capital and that of their friends in local enterprises.

The coastal region was particularly fortunate in attracting the financial attention of John (Bet-a-Million) Gates, of Harris, Gates and Company, an investment firm based in New York and Chicago.[24] Gates came to the area as a result of investments in the KCP&G. The founder of that railroad, Arthur Stilwell, had earlier invested in such varied Port Arthur projects as rice growing, electric power production, and land sales. Gates continued and increased such investments. He became interested enough to build a summer home in Port Arthur. He enticed friends like J. J. Mitchell, president of the Illinois Trust and Savings Bank, and James Hopkins, a successful St. Louis attorney, to join him in Port Arthur investments.[25] Gates was a plunger by nature, and he was a forceful, well-connected man. After he took over the KCP&G, his financial involvement in Port Arthur increased, giving the region a strong personal connection with eastern capitalists. The railroads brought a number of men like Gates to the region. As a source of investment capital and as a major employer and training ground for industrial and administrative skills, the railroads thus exerted a crucial influence on the regional economy before the discovery of oil.

The local economy of the Beaumont-Orange area was not dominated by the railroads to the same extent as that of either Port Arthur or Houston. The Southern Pacific was a large employer, but more important was a business encouraged by the spread of railroads into East

Texas—the lumber industry. Beaumont's initial period of rapid expansion in the 1870s paralleled a post-Civil War boom in the lumber industry in general and in the southeastern United States in particular. East Texas marks the western end of the southern pine belt, and this resource attracted the attention of both local and eastern investors who were looking for new sources of lumber to supplement or replace that obtained from the forests around the Great Lakes.[26] Between 1876 and 1878 four large mills were built in Beaumont. At that time, two Pennsylvania lumbermen, Henry Lutcher and G. Bedell Moore, migrated to Orange and established the first large-scale lumber manufacturing operation in Texas.[27] Lumber was originally cut in East Texas and floated down the Sabine and Neches Rivers to mills at Beaumont and Orange. But in the 1880s and 1890s, railroads replaced the rivers as the primary means of transportation. The Sabine and East Texas, the Gulf-Beaumont and Kansas City, the KCP&G, and a system of tram railways extended from Beaumont and Orange to the long-leaf pine region. As a result, the Sabine-Neches area (including Beaumont and Orange and later Port Arthur) became the largest sawmill center in East Texas.

The existence of large stands of virgin pine, the access to export markets through Gulf ports, and the growing domestic demand for lumber made the East Texas industry a very attractive investment opportunity. All of the easterners interested in investing in Texas did not, of course, follow Lutcher and Moore's example by migrating to the state. Instead, many simply financed the efforts of local entrepreneurs. The early career of John Kirby, who became a major figure in the Texas lumber industry in the early twentieth century, illustrates the benefits of such partnerships. In the 1880s, Kirby, a lawyer in a small town in the middle of the pine belt, impressed Nathaniel Silsbee, a Boston banker, with the prospects for East Texas timber. Silsbee's financial backing enabled Kirby to set up the Texas Land and Lumber Company, which quickly recorded handsome profits. Such results encouraged Silsbee and other Boston capitalists to back Kirby in a larger undertaking, the Texas Pine Land Association, established in 1890. They also supported him in the building of the Gulf-Beaumont and Kansas City Railroad.[28] Other lumbermen did not have the same level of financial backing, but the total amount of capital involved in this industry was nonetheless substantial. In 1900, Beaumont and Orange combined had a population of only about 13,000, but the Census of Manufactures in that year reported almost $6.5 million in capital invested in manufacturing in these two mill towns.[29] The lumber industry encouraged the accumulation of both local and nonlocal investment dollars. Much of this capital ultimately was shifted into other investments in the region when they proved to be more profitable than timber.

Few more lucrative opportunities existed before 1901. A study of the operation of fourteen southeast Texas lumber mills (including five of the largest mills in Beaumont and Orange and nine other mills throughout East Texas) during the period from January 1, 1899, to April 30, 1901, revealed gross earnings of $2,300,000, with expenses and taxes of $1,500,000.[30] The recipients of the resulting profits, the owners of the mills and of the timberland, emerged as the economic elite of the region during the period from 1880–1890. Lumberman ran the only two banks in the area; they manned the boards of trade; they operated the smaller railroads, at least until selling out to the larger systems. These were the men who controlled the area's economy at the end of the nineteenth century.

The experiences of the previous two decades gave them reason for optimism. The Sabine-Neches area had prospered from the expansion of the lumber industry. In 1870, neither Beaumont nor Orange was large enough to merit inclusion in the census of cities and towns. By 1900, they employed more than 2,200 workers in manufacturing and produced more than $2 million in manufactured goods.[31] Sawmills, planing mills, and shingle mills employed most of these workers, and others worked in industries that supplied the mills.[32] Two ironworks in Beaumont provided metal goods for the mills; the Beaumont Iron Works even produced logging cars for the railroads.[33] Several large brickworks furnished the needs of the mills as well as those of the construction industry. Such industries and the cities of Beaumont and Orange would continue to expand as long as the lumber mills prospered.

The future of the coastal region thus was tied through its mills to the future of the East Texas pine region. The rivers and railroads from the coast reached into a hinterland far different from the Sabine-Neches area. There were no towns nearly as large as either Beaumont or Orange in the interior. Instead, small logging centers and railroad settlements supplied the needs of the loggers and the largely self-sufficient farmers who lived among the pines. Beaumont was the banking and commercial center for these East Texas towns.[34] The region's "money crop" was pine trees, and its primary economic activity was the cutting of timber. Trains then brought the lumber to the coastal mills, where it was prepared for sale. Finished lumber was sold in local markets, to the railroads, and to Houston markets. It was also exported via Sabine Pass and Port Arthur to European markets.

By 1900 the lumber industry in Texas, which was confined almost entirely to the eastern part of the state, produced a higher value of product than any other industry. Texas' share in the total value of lumber products produced in the nation had increased to about three percent in 1900. Although this figure is not very impressive on a national scale, the industry was very important on the local and regional levels. Statistics on the region's processing and export of lumber products come from a source that is best treated with skepticism, perhaps as an upper limit. A group of Orange lumbermen presented the following figures to a visiting committee from the New York Chamber of Commerce and Manufacturer's Association. They asserted that, from 1891 to 1900, lumber valued at more than $11 million left Orange, where "the lumber industry supports, directly and indirectly, more than 6,000 people in Orange and vicinity."[35] Beaumont probably surpassed these totals in the same decade.[36] Such extensive exports of lumber products helped create a small yet thriving industrial base in the Sabine-Neches area.

Houston also shared in the growth that accompanied the expansion of timber production. Its prime connection to the timber region was the Houston, East and West Texas Railroad, which supplied timber to several small mills.[37] But the primary economic benefits for Houston from the timber industry did not arise from the milling and shipping of lumber, as was the case with Beaumont and Orange. As the region's center for finance, trade, and transportation, Houston shared in the profits of timber development, even though the lumber was cut and processed elsewhere: "In 1899 the total volume controlled through the Houston lumber market was 429,000,000 feet, most of which never entered the city, but traveled directly over the railroads from East Texas saw mills to the purchaser."[38] Lumbermen like John Kirby and Jesse H. Jones chose Houston as the headquarters for their large lumber businesses. Profits from wood cut in East Texas, milled in Beaumont, and sold in Illinois often accrued to Houston-based companies. Some was reinvested in the lumber industry, but some also found its way into other Houston area enterprises. As with the regional offices of the largest railroads, these large lumbering operations helped create financial, administrative, and communications resources that were then helpful in bringing to Houston the regional offices of other companies in other industries.

Cotton was tied even more closely to the Houston economy than was timber. Cotton had inspired the building of the first railroad near Houston, the Buffalo Bayou, Brazos and Colorado, in the 1850s.[39] The roads linking the cotton markets with the rich blackland prairie to the north and west encouraged the spread and intensification of cotton planting, which in turn fed the further expansion of the railroads. The railroads reduced transportation costs so drastically that even some areas near the Red River often found it profitable to ship by rail to Houston rather than by water to New Orleans.[40] Houston shared with Galveston the role as primary shipping point for Texas cotton, and the preparation of cotton for market was one of the largest manufacturing industries in Houston before the coming of the oil industry.[41]

To supply the raw material for this industry, the blackland prairie cotton fields developed a distinctive agricultural pattern in which almost half of all farm operators were tenants.[42] The cotton belt was very densely populated for a rural area. Harris County, the site of Houston and the focus of much attention in the previous pages, had a population in 1900 of about 64,000 in an area of roughly one million acres. Two adjoining counties in the cotton belt (Falls and Milam) with approximately the same area and with no large cities had a population of more than 73,000.[43] Well over a million people lived in the cotton-rich blackland prairie which stretched to the north and west of the coastal region. This system produced vast quantities of cotton for export, but the large numbers of people involved promised difficult future problems of adjustment. Capital generated from the cotton trade could be absorbed into an industrial economy, as could the administrative and commercial skills developed by traders and transporters in Houston. The absorption of the vast numbers of tenants into an industrial work force would be a far more difficult matter.

In 1900, however, such problems had yet to materialize. Profits in the cotton business were substantial, especially in the region's major city. The cotton came down by rail to Houston, where it was processed or shipped by rail to New Orleans or Galveston for export. As in the lumber industry, Houston-based interests also controlled the sale of some cotton directly from northern Texas to St. Louis. The market for cotton was expanding at the end of the nineteenth century, and Houston's economic role as gatherer, processor, and transporter of that fiber promised continued benefits.

THE REGIONAL ECONOMIC SYSTEM IN 1900 AND AFTER

John Spratt's description of the transformation of the Texas economy in the period from 1875–1901 provides a good summary of developments in the coastal region:

> Within a short generation the wilderness had become an empire. Ten thousand miles of railroad spanned the state. Agriculture had moved far along the road from live-at-home to commercialization. Percentagewise, urban population was increasing much faster than rural. Industry was in a position to challenge the economic leadership of agriculture.[44]

By 1901 the regional economy, spurred by a transportation revolution, was generating significant volumes of timber and cotton for export. The production and handling of these two major products had given rise to a pool of industrial, financial, and administrative skills which, at least in the urban centers, could be utilized for further development.

Timber and cotton each spawned a distinct pattern of economic activity, and the two patterns only partially overlapped. The lumber industry produced a system of small East Texas towns dependent on the coastal cities for markets, for facilities to finish lumber products, and for banking. Compared to other cities throughout the nation, Beaumont and Orange were merely larger versions of the East Texas mill towns, but they served essential urban, industrial, and commercial functions within the region. Houston, which was more than three times larger than Beaumont and Orange combined, had much the same relationship with these cities that they had with the small logging towns in the interior. Superior size and resources and superior connections into the financial and trade centers of the national economy made Houston the primary gateway to the region. In addition, Houston was the heart of a growing system of cotton production that drew the efforts of thousands of small farmers into the national economy. Access to outside markets for both cotton and timber had increased with the expansion of the railroads and the improvement of the Gulf ports of Galveston, Sabine Pass, and Port Arthur. In 1900 both industries were growing and were promoting further urbanization and industrialization in the region.

Some restraint is necessary. Beaumont, Texas, at the turn of the century, was not a mature urban center with tree-lined boulevards and a city symphony. Substantial growth had occurred, but the region was still young. Rough edges remained. Beaumont had few public utilities, no paved streets, no sewer system, and few substantial buildings in 1900. Construction of its first hospital, capable of handling only twenty-four patients, had begun in 1897.[45] Beaumont was less a city than the site of three large lumber mills. Houston had some paved streets, but little else other than a small but aggressive business community. In 1892 one observer described it as follows: "Houston is an overgrown, dirty village, seemingly blundering along without any policy or defined government or management.... Houston is the most dirty, slovenly, go-as-you-please, vagabond appearing city of which I have knowledge."[46] Regional economic development had laid the foundation for further growth but had not yet created the social amenities available in the cities of older, more mature regions.

From the pre-oil perspective of 1900, further regional growth as a supplier of raw materials for the national market seemed assured. The railroads had provided the primary economic impetus for regional growth. But even as the transcontinental railroads came to the region in the 1870s and 1880s, events were underway in other segments of the national economy that would ultimately prove of greater consequence to the region than even the railroads. An expansive sector of the national economy, the petroleum industry, was gradually moving westward in search of crude oil. On January 10, 1901, one spectacular result of this search, the Spindletop gusher, focused the attention of the entire industry and of the nation on Beaumont, a sawmill town in the "boggy country" of southeastern Texas. Thirty years of gradual development had prepared the city and the surrounding region for continued growth, but nothing could have adequately prepared it for the impact of the Spindletop discovery.

NOTES

1. Frederick Law Olmsted, *A Journey Through Texas* (New York, 1857), p. 366.
2. Ibid., p. 65.
3. Kenneth Wheeler, *To Wear a City's Crown: The Beginnings of Urban Growth in Texas, 1836-1865* (Cambridge, Mass., 1968), p. 1. See, also, Andrew Forest Muir, "Railroads Come to Houston,—1857-1861," *Southwestern Historical Quarterly* LXIV (July 1960), pp. 42-63.
4. Dermot Hardy and Ingham Roberts (eds.), *Historical Review of South East Texas*, 2 vols. (Chicago, 1910), p. 198. Marilyn McAdams Sibley, *The Port of Houston* (Austin, 1968) substitutes "350 of 450" for Hardy's "4/5 of 492 miles."
5. John Spratt, *The Road to Spindletop: Economic Change in Texas, 1875-1901* (Dallas, 1955), p. 4.
6. Carl Degler, *The Age of Economic Revolution: 1876-1900* (Glenview, Ill., 1967), p. 24. For a useful description of economic changes in this period, see Edward C. Kirkland, *Industry Comes of Age: Business, Labor, and Public Policy, 1860-1897* (New York, 1961).
7. Julius Grodinsky, *Transcontinental Railroad Strategy, 1869-1893: A Study of Businessmen* (Philadelphia, 1962), chapters 10 and 16.
8. Dermot Hardy, *Historical Review*, p. 241.
9. S. G. Reed, *A History of Texas Railroads* (Houston, 1941) gives the date 1881 for the completion of the T&NO to New Orleans. This date is different in several other sources, including the *Transportation by Land* volume of the 1890 Census, pp. 102-105.
10. Once again, this is Reed's date, and the date differs in other sources.

11. John O. King, *The Early History of the Houston Oil Company, 1901-1908* (Houston, 1959).

12. Julius Grodinsky argues that both the T&NO and a long San Antonio to Houston line were bought at a high price relative to their worth. See Julius Grodinsky, *Transcontinental Railroad Strategy,* p. 169.

13. Keith L. Bryant, Jr., *Arthur E. Stilwell: Promoter with a Hunch* (Nashville, 1971). Stilwell explained his change of plans by revealing that "Brownies" had appeared to him in dreams and advised him to switch sides. The hurricane that destroyed Galveston in 1900 made the Brownies' advice look pretty good.

14. For accounts of the infighting in the rivalry of the promoters of Port Arthur and Sabine Pass, see George M. Craig (Port Arthur) to John W. Gates (New York), letter dated March 31, 1905, "January-June 1905" folder in George M. Craig papers, San Jacinto Historical Museum, San Jacinto Monument, Deer Park, Texas. The rivalry is also described in Keith Bryant's biography of Stilwell. See, also, John Rochelle, "Port Arthur: A History of Its Port to 1963," M.A. thesis, Lamar State College, 1969.

15. For a history of Galveston's early position in Texas, see Earl Fornell, *The Galveston Era: The Texas Crescent on the Eve of Secession* (Austin, 1961).

16. Lynn Alperin, *Custodians of the Coast* (Galveston, 1977), pp. 119-148.

17. Of the eight cities, Yonkers had the greatest population (12,733) in 1870. Excluding Dallas, which was not yet listed by the census in 1870, the other cities all had between 8,000 and 11,000 people.

18. David G. McComb, *Houston: The Bayou City* (Austin, 1969), pp. 109-110.

19. United States Census 1900, *Manufactures,* vol. VII, pp. 880-881. Under the label "cars and general shop construction and repairs by steam railroad companies," for Houston are listed 52 "salaried officials, clerks, etc." earning over $67,000 and nearly 1,300 "wage earners" earning more than three-quarters of a million dollars.

20. R. L. Polk and Co., *Texas State Gazeteer and Business Directory, 1896-1897,* vol. 5 (St. Louis, 1898), p. 221.

21. According to the 1900 census, the average yearly wage in the railroad shops was almost double that of those working in the second largest industrial category, "oil, cottonseed and cake." This is borne out by a comparison of newspaper reports of millhands working for two dollars a day and riceworkers making one and a half dollars, since railroad commission reports show that the railroad machinists averaged more than $3.20 daily. For the statistics for the T&NO, see *Annual Report of the T&NO to the Texas Railroad Commission for the Year Ending June 30, 1901.* Box 4-3/695, Records of the Texas Railroad Commission, State Archives Building, Austin, Texas.

22. For the role of the railroads in American development, see, especially, Alfred D. Chandler, Jr., (ed.), *The Railroads: The Nation's First Big Business* (New York, 1965).

23. The magazine began publication sometime in the 1890s and was distributed by the Southern Pacific Passenger Department. Southern Pacific Passenger Department, *The Coast Country of Texas, 1903* (Orange, Texas, 1903) in the State Archives, Austin, Texas. The Kansas City Southern also published such a magazine.

24. Gates was a man of great energy who began his rise to fortune by selling and manufacturing barbed wire. His nickname is said to have originated on a train trip with a wealthy associate. Bored with the ride, he proposed a million-dollar bet on which rain drop would slide down the train window first. He won, but only after his traveling companion had scaled down the bet. See American Guide Series, *Port Arthur* (Houston, 1940).

25. Gates actually convinced Mitchell to join him in building a summer home in the "city" of Port Arthur. All these men were later prominent in Texaco.

26. See Robert F. Fries, *Empire in Pine: The Story of Lumbering in Wisconsin* (Madison, 1951), and Agnes Larson, *History of the White Pine Industry in Minnesota* (Minneapolis, 1949).

27. Hamilton Easton, "The History of the Texas Lumbering Industry," doctoral dissertation, University of Texas, Austin, 1947, p. 82.

28. King, *The Early History of the Houston Oil Company.*

29. United States Census, 1900, *Manufactures,* vol. III, p. 867. The totals for Beaumont and Orange combined almost equal those reported for the much larger cities of Houston ($6.96 million) and Dallas ($6.9 million).

30. Haskins and Sells (C.P.A. in New York) to Patrick Calhoun (New York), letter dated August 10, 1901, Box 62, C. A. Warner collection, Gulf Coast Historical Archives, University of Houston, Houston, Texas. This study was undertaken as a prelude to the purchase of many of these mills and their consolidation under John Kirby's management.

31. United States Census, 1900, *Manufactures,* vol. III, pp. 868–872.

32. Although exact figures are not available, a publication of the Beaumont Board of Trade lists 300 employees for one mill. There were three or four area mills of approximately the same size and several more only slightly smaller. See Oil Exchange and Board of Trade, *The Advantages and Conditions of Beaumont and Port Arthur Today* (Beaumont, 1902) in the Tyrrell Historical Library, Beaumont, Texas.

33. *Beaumont Journal,* January 8, 1901.

34. The Polk Directory for 1896–97 lists the banking point for each of the small East Texas towns. For an excellent description of working conditions in the lumber industry, see Ruth Allen, *East Texas Lumber Workers, An Economic and Social Picture, 1870–1950* (Austin, 1961).

35. The report is reproduced in Beaumont Oil Exchange and Board of Trade, *Argument Recommending Proposed Improvement of Sabine Lake Channel* (Beaumont, 1902), in Tyrrell Historical Library, Beaumont, Texas.

36. In various business directories, in the county tabulations of the census, and in Beaumont Board of Trade publications, Beaumont is invariably given a sawmill capacity from two to three times that of Orange.

37. *Annual Report of the Texas Railroad Commission, 1901* (Austin, 1901), table 13.

38. David McComb, *Houston,* p. 113, citing the *Houston Daily Post,* April 8, 1900.

39. S. G. Reed, *History of Texas Railroads,* p. 3.

40. C. S. Potts, *Railroad Transportation in Texas,* Bulletin of the University of Texas, No. 119 (Austin, 1909), p. 17. For much more detailed discussions of the railroads' impact on national growth, see Albert Fishlow, *American Railroads and the Transformation of the Antebellum Economy* (Cambridge, 1965). See, also, Robert Fogel, *Railroads and American Economic Growth* (Baltimore, 1964).

41. United States Census, 1900, *Manufactures,* vol. VIII, pp. 880–881. The census reports "oil, cottonseed and cake" as the most highly capitalized (over $1.15 million) and the most valuable (over $1.75 million) manufacturing industry in Houston in 1900.

42. United States Census, 1900, *Agriculture,* vol. V, part 1, tables 10, 19, 55, pp. 125–130.

43. United States Census, 1900, *Population,* vol. I, part 1, pp. 999–1002.

44. John S. Spratt, *The Road to Spindletop,* pp. viii–ix.

45. Maurine Gray, *A History of Medicine in Beaumont* (Beaumont, n.d.).

46. David McComb, *Houston,* p. 137.

Chapter II

The Creation of a Refining Region: 1901–1916

The years from 1901–1916 marked a crucial transition in the development of the upper Texas Gulf Coast. By the end of this brief period, the petroleum industry had begun to impose new economic patterns over those of the pre-oil period. But for several years after the Spindletop gusher, a scramble for wealth similar in tone, if not scope, to the California gold rush placed the regional economy in a state of flux. There was no assurance that the regional impact of Spindletop would survive the life of the oil field, as intense competition among the oil companies in the field eliminated the firms that did not possess adequate financial backing, managerial abilities, or luck. The refineries built during these chaotic early years were initially less important to the regional economy than was oil production and the construction of pipelines. But the extension of pipelines from numerous oil fields discovered in Texas and Louisiana after Spindletop to these coastal refineries gradually increased their importance, and the construction of pipelines from the Port Arthur plants of Gulf Oil and Texaco to the vast oil deposits of the Mid-Continent field in Oklahoma in 1907 assured the continued growth of the Gulf Coast as a major refining center. After the Spindletop boom had subsided and most of the producers had departed for other fields, the refineries of several growing oil companies thus remained as a lasting regional legacy of Spindletop.

SPINDLETOP AND THE OIL INDUSTRY

An understanding of Spindletop's place in the history of the oil industry helps put the region's development in a broader context. The size of the discovery, the oil companies that originated at Spindletop, and the new uses found for its oil all gave it special importance. On a regional level, Spindletop transformed the economic system; on a national level, it helped transform the oil industry.

On the eve of the Spindletop discovery, the Standard Oil Company (New Jersey) dominated all phases of the oil industry in the United States. Although the discovery of new fields in Pennsylvania, Ohio, and West Virginia had helped small independent firms gain niches in the petroleum market, Standard in 1899 still controlled between eighty-five and eighty-eight percent of crude supplies, eighty-two percent of refining capacity, eighty-five percent of the market for kerosene (then the primary refined product), and most of the pipeline and tanker carrying capacity of the United States.[1]

The decade after the Spindletop gusher witnessed the gradual decline of Standard Oil's position. Ralph Andreano has argued convincingly that "the emergence of new competition was a function of the size, quality, and location of new crude discoveries and the limited success in restricting market space in the new fields on the part of the Standard Oil Company."[2] Even before the Supreme Court's dissolution of Standard in 1911, developments in the Gulf Coast, Mid-Continent, Illinois, and California fields had undermined Standard's near monopoly in oil.

Due to its scale, its location near the sea, and the legal environment in Texas, the Gulf Coast field took on a special importance. In the context of the time, the field was very large. In 1902 the total crude production from the two largest eastern fields, Appalachia and Lima-Indiana, was 55.5 million barrels. During the same year, several acres around Spindletop produced eighteen million barrels of crude, twenty percent of the total production in the United States.[3] Such spectacular production from a relatively small field attracted resources that later helped discover and develop even larger fields throughout the Southwest. Because state laws banned the open entry of Standard into Texas, other companies had a chance to grow without directly competing with the dominant firm in the industry. Markets outside of Texas were also within reach, for only a sixteen-mile pipeline over flat ground was needed to give the crude produced by Spindletop-based companies access by tanker to East Coast markets.[4]

Because of these special circumstances, the Spindletop field gave birth to several energetic, expansive new firms. Indeed, Spindletop probably had a greater impact on the market structure of the petroleum industry than has any American oil field discovered in the twentieth century. Gulf and Texaco were born at Spindletop. It was there also that Shell Trading and Transport Company first entered oil transporting on a large scale; its transportation investments later led it into other aspects of the industry. Sun Oil, which had been one of the few competitors of Standard in the 1890s, grew much stronger at Spindletop. The Magnolia Oil Company, absorbed by Socony (now Mobil) in the 1920s, emerged from the new field. During the early years of the field's history, numerous decisions were made that shaped the industry's subsequent market structure. The potential for change is best illustrated by considering the implications for twentieth century oil history of three aborted mergers that were considered during the Spindletop boom. The firms involved were Texaco and Gulf, Gulf and Standard Oil, and Shell and Standard Oil. The energy, the excess, and the opportunity unleased by Spindletop encouraged the growth of oligopoly in an industry previously characterized by near monopoly.[5]

In addition to altering the oil industry's market structure, Spindletop and other new fields encouraged the wider use of petroleum products. With the discovery of large new sources of oil came a search for new uses for oil products. Illuminating oils previously had been the industry's primary product, but after 1901 less expensive crude oil led to the marketing of fuel oil as a replacement for coal. The new oil fields were thus the impetus for far-reaching changes in the energy mix of the nation.[6]

All oil authorities agree on the significance of Spindletop; their differences regarding the level of its importance reflect their divergent historical perspectives. The narrower their focus, the more important Spindletop becomes. Harold Williamson, writing in *The American Petroleum Industry,* asserts that "a new era in domestic crude production began in January 1901, when the famous gusher at Spindletop was struck."[7] Even more enthusiastic is Carl Rister, whose *Oil! Titan of the Southwest* argues that "No other event in the history of petroleum was more important."[8] That would seem to be a sufficient claim to fame for one oil well. But the people of Beaumont went even further. The inscription on the fiftieth anniversary monument at Spindletop modestly reads:

> On this Spot
> On the tenth day of the
> Twentieth Century a New Era
> In Civilization Began[9]

As an evaluation of the discovery's impact on the oil industry, Williamson's view seems most appropriate. But from a regional perspective, the commemorative plaque is not far wrong; Spindletop brought a "new era" of development, if not of "civilization," to the coastal region.

THE REGION IN FLUX

Beaumont was an amazing place during the Spindletop boom. It featured a magnificent gusher spouting oil high into the air, instant millionaires, mobs of self-styled "capitalists," a legendary red light district—in short, all the glamor and excess of an unrestrained rush for wealth. Not as memorable but of more lasting importance than the general spectacle of up to 50,000 adventurers crammed into a saw mill town was the migration of capital, managerial talent, and skilled oil workers into the region. Spindletop pushed the region to a new level of economic activity, thrusting a developing timber region into a temporary role as the center of the twentieth century oil industry. This new role gave the region access to a broad range of developmental resources from the national economy.

The magnitude of these newly available resources is seen in the amount of capital attracted by the boom. In the four years before the discovery of oil, new corporations with Beaumont, Port Arthur, or Orange listed as their principal place of business had an authorized capital of about seven million dollars. The Port Arthur Channel and Dock Company, a corporation financed by the Kansas City Southern Railroad to dredge the ship canal in Lake Sabine, accounted for two million dollars of this authorized capitalization.[10] The five largest lumber companies in Beaumont were capitalized at a total of 1.2 million dollars.[11] The T&NO Railroad, with an authorized capitalization of five million dollars, had the largest capital stock of any corporation active in the region.[12] In the three months after Spindletop, sixty-one new companies chartered at over twenty-four million dollars claimed Beaumont, Orange, or Port Arthur as their place of business.[13] By the end of 1902, as many as 400 oil companies chartered at about $175 million had been created in response to Spindletop.[14] A more telling figure is the amount of money actually invested in oil development. For the period from January 1901 to September 1904, a careful survey by the *Oil Investors' Journal* estimated that about forty million dollars had been invested in Southeast Texas in the drilling, producing, transporting, and refining of oil.[15]

This capital came from within the region, from all over the nation, and even from Europe. In discussing the failure of a large company supported primarily by capitalists from St. Louis, the *Oil Investors' Journal* reported that "nearly everyone of our foremost cities has had its 'local' Beaumont oil enterprise. It is not a hard matter for promoters to interest good men in the big towns and to enthuse them."[16] The names chosen for various companies, Boston Beaumont Oil Company, Baltimore and Beaumont Oil Company, Iowa-Beaumont Oil, Indiana Oil Company, New York Oil and Pipe Line Company, and various other geographical designations, indicate the widespread national participation in the Spindletop boom. These dollars encouraged the development of the region as well as the growth of the individual oil companies.

The fact that some reporters labeled the new field "Swindletop" suggests that not all of these investments were productive.[17] Fake companies and get-rich-quick stock schemes abounded in the boom days. Widespread fraud meant that much of the nearly $200 million in authorized capital stock never really existed or at least never found its way into the development of the oil field. Many additional investment dollars were lost when legitimate oil companies went bankrupt. Intense competition for oil was just as prevalent as the intense competition for the money of the unwary investor. Many fortunes, but only a few permanent oil companies, were built during the Spindletop boom.

A study of investments in individual companies suggests no simple patterns. Capital came from men as different in station yet as similar in outlook as "Andrew W. Mellon, Pittsburgh, Pennsylvania, Capitalist" and "Howard Bland, Taylor, Texas, Capitalist." One result, however, does stand out. Those companies that succeeded were usually operated by experienced oil men from existing fields who had made strong, early, and continuing contact with large eastern investors. The profits from the operation of a 200-employee saw mill were not sufficient to finance an integrated oil company. The region had moved to a new scale of investment, and in the United States at the turn of the century, large sources of capital were concentrated primarily in the East.

The best connected "local" oil company was the original discoverer of Spindletop, the J. M. Guffey Petroleum Company. During the 1890s two Beaumonters, Patillo Higgins and Anthony Lucas, believed that a salt dome near Beaumont contained oil, but they could not persuade either local investors or Standard Oil to provide capital for exploration. Support finally came from James Guffey and John Galey, a Pittsburgh-based oil prospecting partnership. Guffey and Galey arranged a loan of $300,000 from T. Mellon and Sons, a Pittsburgh banking house operated by the Mellon family, which had accumulated its wealth in railroads, steel, oil, and banking.[18] The partnership between Guffey and the Mellons was strained from the start, and the Mellon interest gradually gained control of the company. The original Guffey company included on its board of directors a powerful group of Pittsburgh men—Andrew Mellon, future Secretary of the Treasury; Henry Clay Frick, former chairman of the executive committee at Carnegie Steel; James H. Reed, law partner of U.S. Attorney General Philander C. Knox; and T. Hart Given, a prominent Pittsburgh capitalist.[19] Involved in the company at its founding or shortly thereafter through bond issues were others who had been prominent in Carnegie Steel, which had only recently been purchased by the U.S. Steel Corporation. The backing of men like Frick, Charles M. Schwab, and Joshua Rhodes (a former president of the National Tube Company, the leading producers of pipe in the United States) gave the new corporation a degree of financial stability unusual in the turbulent oil business, at least outside the Standard Oil domain. The inclusion of others like William Flinn, "boss" of Pittsburgh politics, gave the Guffy Company an additional dimension of power and influence.[20] Such investors helped make the company, which was renamed Gulf Oil in 1907 after Guffey's departure, a powerful force in the new oil field, in the oil industry, and ultimately in the national economy.

Although no other local oil company could match Gulf Oil's powerful connections, many had access to eastern capital. The Texas Fuel Oil Company began with a capital stock of $50,000 supplied primarily by experienced oil men who had come to Beaumont from the eastern fields by way of a small refinery in Corsicana, Texas. The success of the company and its needs for additional capital with which to expand led to the creation in 1902 of Texaco, with an authorized capitalization of three million dollars. Investment in Texaco came primarily from two sources: Arnold Schlaet, who had been active in the eastern oil fields, and who secured backing from his associates John J. and Lewis Lapham, large wholesale leather dealers in New York; and John W. Gates of Harris, Gates and Company. Both sources of capital proved to be generous and patient. Gates, who had earlier ties with the Kansas City Southern Railroad and a financial interest in the growth of Port Arthur, provided substantial and sustained support, and he became a dominant figure in the early history of Texaco. His enthusiasm also persuaded several of his wealthy business associates to invest in the new corporation. Texaco, originally more truly a local company than Gulf Oil, also drew financial backing from Texas sources that included James S. Hogg, former trustbusting governor of Texas, and James Swain, Hogg's former floor leader in the Texas House. Such local investors, combined with Texaco's strong ties to eastern sources of capital, gave it excellent financing.[21]

The most ambitious combination of local and outside capital involved the operations of John Kirby, the Houston-based lumberman. Kirby responded quickly to the discovery of oil. As a lumberman he owned considerable land in East Texas, and he felt that this made his lumber business a natural partner for a venture in oil exploration. His previously established contacts in the East aided in the chartering of the Beaumont Oil Company (one million dollars authorized capitalization), with several other local lumber men and George Silsbee of Boston as participants.[22] But Kirby had much more gradiose visions. He interested Patrick Calhoun, the grandson of John C. Calhoun and a corporate attorney in New York, in plans for a giant cooperative enterprise in lumber and oil.[23] Calhoun obtained expert information on the feasibility of such a corporation, and after a glowing report that asserted that oil would be found throughout the timber region, he approved the deal. With the backing of Brown Brothers and Company, Simon Borg and Company, and the Maryland Trust Company, two new corporations, the Houston Oil Company (thirty million dollars) and the Kirby Lumber Company (ten million dollars) were chartered on July 1, 1901. Houston Oil was to buy vast tracts of timber lands and help Kirby to acquire fourteen existing timber mills. Then Kirby's annual stumpage fees (money paid to a land owner for the right to cut his timber) of about $500,000 per year was to provide the working capital for oil exploration on the timber lands. Financial disputes foiled this promising arrangement, but the venture nevertheless produced a major lumber company and a major oil enterprise.[24]

Capital flowed into the region from many other sources. Despite the fact that Standard Oil was barred by law from entering the field, the Seaboard National Bank in New York financed the three million dollar Security Oil Company in the interest of the industry giant.[25] The Lone Star and Crescent Oil Company, with a chartered capitalization of ten million dollars, attracted substantial investment from New Orleans capitalists.[26] A former U.S. senator from Minnesota headed two corporations with authorized capital of $4.5 million.[27] During 1901 and 1902, forty-four oil companies chartered to operate in the Beaumont area each had over one million dollars in authorized capitalization. Those discussed above were the most successful, not necessarily the best financed. Investment funds flowed into the oil industry both from national centers of industrial power and from local citizens. With all of this money seeking the same oil, a second resource, managerial talent, became extremely important.

The Pennsylvania, Ohio, and West Virginia oil fields had been the training schools for most of the oil men who migrated to Texas, and Standard Oil had been their primary instructor. The Spindletop discovery drew these men to Beaumont, and they were the most important force working to create an orderly pattern of development. The evolution of the region into a refining center resulted from the ability of these experienced oil men to organize integrated oil companies designed after the existing model for successful oil operations, Standard Oil.

Skilled managers and experienced petroleum specialists were especially important in the dramatic rise of Texaco, and a detailed examination of its personnel helps explain both the firm's growth and the general process by which experienced oil men came to the region. The most important Texaco officer and one of the most important figures in the American oil industry of the early twentieth century was Joseph S. Cullinan. After a childhood in Oil Creek at the center of the Pennsylvania oil region, Cullinan worked his way up from the bottom at Standard Oil between 1882 and 1895. He rose to a position as a division superintendent of Southwestern Pennsylvania Pipe Line Company, a Standard subsidiary, before resigning to help establish the Petroleum Iron Works, a supplier of various metals products to Standard and other companies. In 1897 he came to Texas to operate the first refinery in the state, a small plant at Corsicana (south of Dallas) that was financed by Calvin Payne and Henry Folger of Standard Oil. Cullinan's outstanding reputation and experience were the primary assets with which the small Texas Fuel Oil Company attracted the large investors that enabled it to prosper as the Texas Company.

Many of those working under Cullinan at Texaco shared a similar past. The second in command was Arnold Schlaet, whose involvement with the Elcho Oil Company in Pennsylvania had trained him in the financial and marketing aspects of the industry. To lay its pipelines, Texaco called on W. J. Ennis, a pipeline builder well-known in the Pennsylvania fields. When R. C. Holmes, who had worked for Standard Oil in Pennsylvania and Ohio, supervised the construction of the original Texaco refinery, he built from a design provided by W. T. Leman, an experienced refiner who had been the supervisor of the Manhattan Oil Company, a Standard affiliate, in Welker, Ohio, and of the Lake Carrier's Oil Company in Coraopolis, Pennsylvania. Following the completion of the plant in 1904, Leman became head of Texaco's refining department.

Leman's migration to Texas illustrates the informal system of friendships and working relationships that attracted oil specialists to the region and directed them into specific companies. While in charge of the Welker, Ohio, refinery, Leman's subordinates included C. C. Blackman, a foreman, and J. C. McCullough, the superintendent of construction. Both preceded him to Texaco.[28] Their prosperity and the promise of a healthy salary increase encouraged their former boss to move from the declining fields in the East to those in the rapidly expanding Gulf Coast region. Cullinan made similar use of his former associates in the East in recruiting other skilled oil men. In one letter to a Pennsylvania friend, Cullinan asked for information about "the best all around comer in the pipeline or oil business that you know of in the North." The criteria suggested for choosing the proper man—"good health, strength, and intelligence with some practical knowledge and a considerable amount of horse sense"—aptly describe many of the men drawn together by Cullinan at Texaco.[29]

Other companies cultivated similar connections with experienced oil men in the eastern fields. Gulf's first manager, J. D. McDowell, had spent eighteen years at Standard Oil as an engineer before becoming involved in 1899 with the Mellons, who dispatched him to Beaumont in 1901. When his performance proved unsatisfactory, Andrew and Richard Mellon sent down their nephew, William Larimer Mellon, who had organized and operated a successful West Virginia-Pennsylvania oil producing, refining, and pipeline company before he sold out to Standard Oil in 1895 at the age of twenty-seven. One of Mellon's first acts was to hire George Taber away from the Atlantic Refining Company of Philadelphia, the largest refining company controlled by Standard, and make him the head of Gulf's refinery.[30] Security Oil Company, which had financial connections with Standard, not surprisingly also drew upon that company for employees. The builder, the superintendent, and the first two managers at Security's refinery had worked for Standard Oil affiliates immediately before coming to Beaumont.[31]

Not all men with the skills needed to be managers, drillers, or refiners had acquired their knowledge as Standard employees. J. Edgar Pew and Sun Oil, one of the largest eastern independents in the late nineteenth century, prospered at Beaumont. The Heywood Oil Company parlayed the luck and drilling talents of three brothers who had not worked for Standard into a prosperous company. What was important was some kind of practical experience, and in the last thirty years of the nineteenth century most, but not all, of those acquiring such experience had at one time or another worked for the Standard Oil Company.

For administrative, legal, and marketing skills, the companies drew heavily from the local economy. The railroads were one source of such personnel. Gulf hired away a general manager from the Southern Pacific Railroad,[32] and Texaco reached into the same company to hire an engineer to oversee the construction of a pipe line from Oklahoma to Port Arthur.[33] Most of the companies hired lawyers from the region or at least from Texas. Accountants, purchasing agents, and salesmen generally came from local industries and from the government.

Given the pace of their early development, the regional oil companies could not expect to attract skilled workers in sufficient quantities. They had to train many employees. Along with experienced oil workers, thousands of unskilled men came to Spindletop. Most of this potential work force came from the surrounding areas in Texas and Louisiana and their motive was simple: "I just felt there might be a good opportunity for a young man to get work."[34] Those who had come to Spindletop with previous experience quickly trained this work force. In turn, as other oil fields opened up in the Southwest, Spindletop workers supplied much of the knowledge and experience necessary for their development.

Of course, this process of migration was not always orderly. The spectacular opportunities promised by southwestern oil brought an immediate and almost overwhelming response from a national economy characterized by good transportation and communication systems and extremely mobile supplies of capital and labor. Oil workers in eastern fields were willing and able to move to a previously unheard of place in southeast Texas seeking advancement and adventure. Investors throughout the nation showed an incredible eagerness to sink their money into ventures such as the Rock Rim Oil and Transport Company in an effort to get in on the ground floor of a major new oil company. Most invested their dollars or their labors in companies that failed; but some correctly chose Texaco or Gulf amid the considerable confusion of the boom period.

From a combination of these migrating resources with those already in the region, individual oil companies sought to mold efficient organizations. Those that succeeded gradually brought order and development, and in the process they also planted the seeds that would grow into the giant Gulf Coast refining complex. As the output of the Spindletop field declined, much of the capital and labor it had attracted moved on to other boom towns throughout the southwestern oil fields. But a crucial difference gave Beaumont's oil bonanza a more lasting impact: the timing of the Spindletop discovery and its location near the Gulf of Mexico encouraged the emergence of a refining complex that became a permanent connection between the region and several large, growing oil companies.

THE EMERGENCE OF A REFINING CENTER

Two related processes helped to create a refining region from the resources drawn by the Spindletop discovery. First, fierce competition eliminated most of the individual oil companies operating in the field.

The second process occurred within the companies that survived. Gulf, Texaco, and, to a lesser degree, several other major regional firms grew into powerful, vertically integrated companies with extensive operations in the producing, transporting, refining, and marketing of oil. Their search for new sources of crude and for wider markets led these enterprises out of the region, but their growth in other phases of the oil industry encouraged the expansion of their refineries on the Gulf of Mexico. By the eve of World War I, major refineries of these large integrated oil companies had become a central part of the coastal region's economy.

With the historian's advantage of hindsight, these developments seem simple and almost inevitable. But, in fact, they resulted from a complex, unpredictable process whereby specific oil companies adopted strategies of integration that, in turn, determined the size and the location of individual refineries. There was nothing inevitable about the growth of Texaco or Gulf; there was no assurance that the dominant companies would build large, permanent refineries in the area around Beaumont. It was not even certain that Texas crude, with its high sulphur content, would be particularly valuable in the long run. The emergence of a regional refining complex resulted from the fact that those few companies that ultimately "won" in the Texas coastal fields had originally chosen Port Arthur and Beaumont for refinery sites.

Gulf Oil (Guffey) was one of the surest bets to win from the start. Its excellent financing and its control of much of the original Spindletop field meant that only monumental mistakes could remove Gulf as a dominant company in the coastal economy. In its early years, however, Gulf *did* make several near fatal mistakes. First, it failed to secure control of the area immediately surrounding the original gusher. Then it sold a valuable strip of land at Spindletop to a syndicate headed by Hogg and Swain, who promptly resold the land in tiny strips to numerous small producers. As a result, Gulf lost control of much of the production of Spindletop. Gulf's other blunder was even more dangerous. In the euphoric boom days, it made a long term contract to supply Shell Transport and Trading Company with large quantities of inexpensive oil. After production later declined at Spindletop and oil prices rose dramatically, Gulf could no longer fulfill its commitment. Had Shell insisted on enforcing the contract, Gulf would surely have collapsed. At least one other large, promising firm, Lone Star and Crescent, went into receivership because of such a contract. But Andrew Mellon's persuasiveness and Shell's desire to cultivate good relations with a large supplier of Texas oil resulted in the renegotiation of Gulf's contract, thereby saving the firm. These early problems and the difficulties of supervising operations from Pittsburgh convinced the Mellons that they should get out of the oil business. The only available buyer for such a vast enterprise, however, was Standard Oil, and it refused Gulf's offer in 1902. In order to salvage its investment, the company was thus forced to pursue a vigorous policy of expansion and vertical integration.[35]

Even before the Standard Oil negotiations, Gulf had invested large sums in pipelines to transport oil to Port Arthur, where it was shipped in tankers to eastern markets. These investments were essential for a company that produced large quantities of crude oil because sufficient markets for this crude did not yet exist in the Southwest. The need to sell oil in large quantities also led to the construction of Gulf's first refinery, a small plant that removed the sulphur from Spindletop crude to increase its value as fuel oil. The choice of a refinery site was dictated by the market for treated oil. From the beginning, Shell was Gulf's largest customer. Its contract with Gulf for unlimited quantities of oil led Shell to develop a large fleet of tankers to carry oil from Texas to European markets, and a Gulf Coast refining location was thus attractive.[36] Port Arthur, with an inland location protected from storms, a ship canal to the Gulf of Mexico, and large vacant tracts of land available at low prices, was chosen. To the initial small refinery was quickly added a large plant to remove sulphur from the Spindletop oil. Then in 1902 Gulf constructed a much larger refinery capable of producing kerosene. With the completion of this plant, it began to compete with Standard Oil for markets in Florida, Louisiana, and even in the Northeast.

Gulf thus emerged relatively quickly as a large refiner. As of 1905, its Port Arthur plant had cost almost two million dollars and had a capacity of about 12,000 barrels of crude per day. This represented only approximately 5 to 6 percent of Standard's total capacity, but the industry giant had eighteen major refineries. Gulf's daily refined output was larger than that of all but five of the existing Standard plants and larger than that of any other independent refinery.[37] To keep a facility of this size operating at near capacity, Gulf vigorously sought new sources of crude. As fields opened up at Sour Lake, Batson, Humble, and other areas to the west of Beaumont, it extended pipelines out from Port Arthur. After the discovery of the vast new Glenn Pool of high quality crude in Oklahoma in 1906, Gulf constructed a 450 mile long, eight inch pipeline from there to Port Arthur. The process of integration had come full circle. A small refinery built to remove sulphur from crude had grown large enough to dictate the commitment of substantial resources to a transportation system to bring crude to Port Arthur. The delivery of the first oil through this line in October 1907 assured the further rapid expansion of Gulf's Port Arthur works.[38]

During the next decade, this plant grew steadily and at times spectacularly. By 1916 a total investment of approximately $5,000,000 had raised its daily capacity to about 50,000 barrels. (See Appendix 1.) In only fifteen years this refinery had become one of the three largest refineries in the nation, rivaling in size Standard of Indiana's plant at Whiting, Indiana, and Standard of California's refinery at El Segundo, California. In 1911 Gulf built a second, much smaller refinery at Fort Worth, but throughout the early period, the Port Arthur works remained by far both the largest refinery owned by Gulf Oil and the largest operated by any company on the Gulf Coast.[39]

Texaco did not have sufficient financial resources to expand its refining capacity as quickly as did Gulf. Its primary energies throughout the early period were devoted to the production of large quantities of crude. An early gamble in acquiring a one million dollar lease at Sour Lake, an unexplored region near Spindletop, caused Arnold Schlaet to lament that "if we should take Sour Lake and it should turn out to be not good enough, that would practically wind up the Texas Company."[40] But the lease proved more than "good enough," and it gave Texaco a solid foundation in crude production from which to move gradually into refining.

Despite this early success, Texaco was still not able to operate independently of Standard as quickly as its better financed rival, Gulf. Both companies cultivated southwestern fuel oil markets with the railroads, Louisiana sugar plantations, and other local industries, but Texaco's primary early buyer was Standard Oil.[41] Arnold Schlaet, who established his New York office "a stone's throw from 26" (Texaco's code name for Standard), dealt extensively and on a personal basis with Henry Fisher, an assistant manager of Standard's fuel department.[42] Standard Oil tankers initially carried most of Texaco's oil to the East Coast. When Texaco decided to construct a refinery, Standard, through Fisher, extracted a promise that Texaco's refined products would be offered to Standard before other markets were sought.[43] Cullinan felt that the expansion of refining in 1904 was "undoubtedly the most important step that we have taken since organizing the company." To assure a market for its refined products, Texaco sent early samples to Standard for approval and suggestions for improvement.[44] While this link with Standard was initially beneficial, Texaco's management recognized that the company's growth called for more independence. In 1904 Cullinan correctly assessed the performance of Standard Oil in the Gulf Coast field in a letter to Schlaet: "They (Standard) are running a kind of incubator, fostering local competition, which in turn will assure their getting the supplies wanted at a very nominal margin as between cost of production and delivery to them, and this is a feature that we should aim to correct, if we undertake to expand and handle the business on broader lines."[45] By marketing in the Southwest, by producing in new fields, by expanding its pipelines from Port Arthur to fields in East Texas and northwest Louisiana, and by a developing its refining capacity, Texaco gradually decreased its dependence on Standard Oil.

The location of Texaco's refinery at Port Arthur reflected its need for a Gulf port accessible to the large tankers of the Standard Oil Company. The choice of Port Arthur over Sabine Pass resulted from Gates' interest in the development of Port Arthur. The company's initial emphasis on the sale of crude oil meant that its refinery expanded slowly. By 1905 it had a daily capacity of only 2,000 barrels a day and represented an investment of over $675,000. After the collapse of plans for a joint pipeline with Gulf to the Oklahoma fields, Texaco spent nearly $6,000,000 in the construction of a 560-mile-long pipeline eight inches in diameter from Port Arthur northward to the Glenn Pool.[46] This investment reflected a major commitment to attaining a better balance between producing and refining.

The subsequent expansion of the Port Arthur works fulfilled this commitment. Its daily capacity grew gradually, reaching 7,500 barrels in 1912. In the next four years a massive construction program increased this figure to 32,000 barrels per day. Earlier, in 1908, Texaco had built a second refinery near Dallas. By 1916 it was clear, however, that Texaco's northern Texas plant, like that of Gulf Oil, would remain much smaller than its Gulf Coast counterpart. At that time Texaco's Port Arthur refinery had grown from its modest beginnings in 1903 into a plant that was large even by national standards. While it was still only about half as large as its neighbor on the flat marshlands southwest of Port Arthur, the Gulf Oil refinery, the continued prosperity of its owner augured well for its future expansion.[47]

The third major refining company, Security, established its plant some fifteen miles away from Port Arthur in Beaumont. In 1902 a mysterious stranger, one George Burt, got off a train, bought a large tract of land on the Neches River near the Spindletop field, constructed a high fence around the land, and built a large, well-equipped refinery. The refinery was paid for by people closely connected to Standard Oil, it was operated by former employees of that firm, and it sold its products almost exclusively to Standard. It was, in effect, a subsidiary of Standard. Unlike Gulf Oil or Texaco, Security initially directly owned no producing facilities. It simply bought oil from others, treated it, and shipped it in railroad tank cars or Standard-owned tankers to other Standard affiliates.

The location of the Security refinery at Beaumont instead of Port Arthur was dictated by the necessity of gathering oil from various sources near Spindletop and by the adequacy of rail shipment for most of Security's transportation needs. To ship oil by water, Security built a pipeline to Sabine Pass. By 1905 the Security refinery had a capacity of about 4,000 barrel per day, which represented an investment of over $2.7 million.[48]

At this point, however, matters of political economy intervened to thwart Security's further development. The prosecution of a state antitrust case against the company in 1907 made it difficult for Security to take advantage of the new Glenn Pool. While Texas was trying its case against Security, the firm transferred most of its men and some of its physical plant to Baton Rouge, Louisiana. In 1909, the Beaumont plant and another in Corsicana were confiscated as illegal Standard Oil affiliates and sold at auction by the state. In response, Standard built a pipeline from Oklahoma to its refinery in Baton Rouge, two hundred miles east of Port Arthur in a state with a less hostile antitrust law than Texas.[49]

Standard's Baton Rouge works was the first major Gulf Coast refinery outside of the area around Spindletop, and the record of its subsequent growth reads like a fisherman's tale about the one that got away. In 1911 the U.S. Supreme Court dismembered Standard Oil, separating its unified holdings into numerous smaller companies. One of the companies thus created, Standard Oil (New Jersey), emerged after the court's action in urgent need of increased refining capacity, and it invested heavily in the Baton Rouge plant in order to meet a portion of its needs. Already by 1916 the refinery had expanded to a capacity of 20,000 barrels per day, and it later became for many years the largest refinery in the world. Throughout the early period, it remained the only major Gulf Coast refinery outside Jefferson County; and for the rest of the century, it remained the largest center of Gulf Coast refining outside of the upper Texas Gulf Coast.[50]

The Beaumont refinery abandoned by Standard's affiliate, the Security Oil Company, also prospered after the state ordered its sale. It was purchased by the Magnolia Petroleum Company, which enjoyed the financial backing of men prominent in Standard Oil (New York). During portions of the antitrust trial, the Beaumont plant had been shut down, and several troubled years of partial operation followed its take-over by Magnolia. The new company gradually overcame these difficulties, and from 1914 to 1916 the plant expanded steadily to a capacity of approximately 25,000 barrels per day. Throughout this period, Magnolia retained strong working ties with Standard of New York (later renamed Mobil Oil), and these ties were finally made formal by merger in 1925. Despite the legal problems of its early history, Magnolia's Beaumont works gradually reached parity in size with the other major Jefferson County refineries, those of Gulf and Texaco at Port Arthur.

In the period immediately after the Spindletop discovery, several firms in addition to Gulf, Texaco, and Magnolia sought to build and operate regional refineries, but none prospered over the long-run. The most substantial refining enterprise that failed was the Central Asphalt and Refining Company, whose plant was second in size only to Gulf's for several years. The refinery, located between Beaumont and Port Arthur on the Neches River, was designed to produce primarily asphalt, but financial and legal problems plagued the original owners and the plant actually refined very little oil. General Refining Company purchased the facility in 1904 and hired R. C. Holmes away from Texaco to be superintendent. In 1907 Texaco reacquired Holmes by buying the 2,800-barrel-per-day General refinery, which it used to produce asphalts in combination with the lighter products made at Port Arthur. Other failures were not salvaged. The Forward Reduction Company had good financial backing and a substantial physical plant, but the failure of its secret refining process reduced it to bankruptcy. Numerous refineries shared a similar fate.[51]

The ability to integrate vertically was one of the most important determinants of success or failure for these early refining companies. Refining was the last branch of the industry entered by all the largest Gulf Coast oil companies except Security. Two, Higgins and Heywood, never built refineries and did not survive as major companies. Gulf and Texaco used the profits and the collateral provided by the production and marketing of large quantities of crude oil to establish and then gradually expand their plants. The possibility of making higher profits from the sale of finished products encouraged their entry into refining. But both also realized that, in the long run, only by duplicating Standard's integrated structure could they survive in competition with the industry giant. This was especially true for Texaco, which could not hope to prosper indefinitely as a supplier of crude to a buyer as powerful and as fickle as Standard Oil. For Texaco and Gulf, pipelines to Oklahoma were firm declarations of a determination to become major integrated oil companies capable of operating independently of Standard. The Gulf Coast refineries at the ends of these pipelines became central parts of the operations of the two largest oil companies that emerged from the burgeoning southwestern oil fields. The coastal region's historical ties to Gulf Oil and Texaco proved to be most beneficial to the regional economy as these two companies grew into giant multinational firms.

Most of the direct economic benefits of the early growth of refining were thus concentrated in Beaumont and Port Arthur. (See Table 2.1.) As of 1916 Jefferson County, which contained both cities and the large refineries owned by Gulf, Texaco, and Magnolia, had almost eighty percent of the total refining capacity located on the Gulf Coast. Standard's Baton Rouge plant accounted for most of the remaining twenty percent. Led by the Gulf Oil refinery at Port Arthur, Jefferson County alone produced over eleven percent of the nation's refined petroleum products in 1916, and it would continue to account for ten to fifteen percent of the national total throughout most of the century. Despite its continued expansion, however, the rapid rise of Houston area refining after World War I meant that Jefferson County would never again dominate regional refining statistics as it did from 1901 to 1916.

During this early period, the overall growth of Gulf Coast refining brought a marked shift in the geographical distribution of national refining. From no capacity in 1901 the Gulf Coast came to provide about fifteen percent of national refining capacity in 1916. In roughly the same period, the percentage share of the two leading eastern refining states, New Jersey and Pennsylvania, plummeted from fifty-four to twenty-eight. Up until World War I, California's refining capacity grew even faster than that of the Gulf Coast, but this would later change as the shift of refinery construction to the Gulf Coast accelerated in later years.[52] Within the national economy, the coastal region was thus gaining a functional identity—that of a major refining center with access to East

Table 2.1. Operating Refining Capacity on the Gulf Coast: 1904-1916 (in 42 gallon barrels per day)

	1904	1909	1914	1916
Total capacity	15,000	38,000	91,000	154,000
Jefferson County	14,000 (98%)	32,000 (84%)	68,000 (76%)	120,000 (78)
Houston area	200 (1)	1,000 (3)	3,000 (3)	3,000 (2)
Louisiana	200 (1)	5,000 (13)	20,000 (22)	31,000 (20)

Source: See Appendix 1.

Coast markets. Even in the early period this proved to be a prosperous calling, as the amount of refined petroleum products consumed almost quadrupled from 1899 to 1914.[53] As the emerging refining region shared in the fruits of this expansion, its role as refiner, producer, and transporter of petroleum products greatly influenced its overall economic development.

THE OLD REGION AND THE NEW

The migration of a large, dynamic manufacturing industry into a region previously characterized by much smaller, timber-related manufacturing quickly altered the pace and direction of regional development. The rapidly growing oil industry spawned a distinctive new economic system based on the production, shipping, and refining of petroleum products. But oil-induced growth did not destroy other industries in the region. Instead, during the general expansion brought about by the oil boom, the railroads, timber, cotton, and rice continued to prosper as secondary industries.

Oil's impact on the regional transportation system exhibited several general characteristics typical of its influence on the region as a whole. The railroads, which provided the initial link between the oil fields and markets for crude, were greated affected by the discovery of oil. Shipment of crude in tank cars enabled companies to market their oil while building pipelines and acquiring access to tankers. But even after these other forms of transportation were available, the railroads remained an important link between the oil companies and southwestern markets for crude. The railroads' response to oil included the extension of existing lines to newly discovered fields and the expansion of tank car capacity.[54] Oil's impact on the region's shipping facilities was even more dramatic. The demands of the new industry led to a vast expansion of existing docks and loading facilities at both Port Arthur and Sabine Pass, as the major oil companies built private docks and wharfs for the handling of oil products. To these docks came a wide variety of tankers owned by Shell and Standard Oil, and, in addition, Gulf and Texaco gradually built fleets of their own. The pipelines leading to the docks and to the refineries were the major innovation in transportation introduced by oil, and they gradually replaced the railroads as the primary regional transportation network for oil. The growth of the oil industry also encouraged the construction of intraurban electric trains in both Port Arthur and Beaumont, of an interurban line connecting the two cities, and of numerous paved roads within the area. The early years of oil industry expansion thus encouraged improvements in regional transportation. Much of the resulting system—the railroad tank cars, petroleum pipelines, and private terminals for shipping petroleum products—was used exclusively by the oil companies and did not directly benefit those outside of the oil industry. Other industries and the region as a whole, however, shared some of the benefits of the more general expansion in transportation facilities—paved highways, intraurban electric trains, and deepened shipping lanes—encouraged by the growing oil industry.

Other local industries responded to the stimulus afforded by the oil industry in ways similar to the railroads. Although exports of lumber from the region remained relatively stable during the boom period after Spindletop, the construction of oil rigs and refineries and the residential and business construction induced by oil-related migration into the region called forth an increased output from local timber mills.[55] Investment by local lumber men in oil companies was also widespread. Several formed successful producing companies that supplied oil to the major firms. But none of these efforts was as important in the development of the lumber industry as John Kirby's enterprise. In the years preceding the Spindletop discovery, the East Texas lumber industry had matured to the point where the consolidation of small mills and land holdings into a large concern promised to make the regional lumber industry competitive in national and world markets. The Houston Oil Company-Kirby Lumber Company deal, despite its failure as a joint enterprise, provided the impetus for this consolidation. The Kirby mills had a potential combined capacity of 950,000 board feet per day, almost ten times the capacity of the largest regional lumber company before Spindletop. With this level of output, Kirby could compete for the large orders that had earlier been beyond the capabilities of local mills. In the period from 1900 to World War I, both production and market concentration in the Texas lumber industry peaked, and the Kirby Lumber Company, an off-spring of the oil industry, was the major Texas lumbering concern during these years.[56]

Oil's impact on the rice industry was similar, though on a smaller scale, to its effect on timber. Rice had been introduced into the region in the late 1890s, and many of its growers feared that oil would cripple their young industry by driving up land prices and by diverting capital and labor from rice. These fears proved false; indeed, rice benefited from the general prosperity encouraged by the oil industry. Capital and labor originally attracted to the region by oil eventually found their way into the rice industry. The growth of a larger local market increased the demand for this foodstuff. In 1902 fourteen new rice companies organized in the region had a capitalization of almost two million dollars. In only two years after 1900, the amount of land devoted to rice growing around Port Arthur approximately quadrupled. In 1907 the area around Beaumont was growing more than two million bushels of rice, four to five times as much as it had in 1900.[57] Although no giant combinations along the line of Kirby's timber and oil venture directly tied rice and oil, several investors, including John Gates, had large stakes in each industry. Gates' mill, like several others in the area, would probably have grown in the absence of oil, but the oil boom accelerated the development of this agricultural industry.

Despite the fact that both their products and their skills were useful to an industry that had large requirements for iron, local iron works benefited less from the oil boom than did other existing industries. The Neches Iron Works in Beaumont expanded immediately after Spindletop, but it had declined slightly by 1906. The Beaumont Iron Works was about the same size in 1906 that it had been in 1900.[58] Neither concern grew very much because of the specialized nature of the oil industry's demand. As they had done in obtaining capital, managerial talent, and skilled labor, the oil companies looked to established suppliers in the older fields for most of their equipment. The managers of Gulf, Texaco, and Security all made extensive purchases from iron works they had used during their previous experiences in the eastern oil fields. Many of these national supply firms established temporary branch offices in Beaumont in order to maintain close contact with their markets, but none moved their major office or manufacturing plant to the Spindletop area.

The fate of the local iron works serves as an important reminder: the oil industry drew from the existing local economy, but its primary source of men, dollars, and equipment was the pool of resources and industries which served the national petroleum industry. Oil influenced regional growth indirectly by encouraging existing local industries, but the primary impetus for growth was the expansion of the oil industry itself.

The pattern of population growth within the coastal region reflected the impact of petroleum-led expansion. From 1900 to 1920 the fastest growing county on the upper Texas Gulf Coast was Jefferson County, which expanded from less than 15,000 to more than 73,000 people. In the same period, Beaumont's population more than quadrupled, and the refinery "town" of Port Arthur grew from 900 to 22,000. (See Table 2.2.) In 1920 these two Jefferson County cities combined were about fory-five percent as large as Houston, a percentage they had never previously exceeded and would never again match.

Table 2.2. The Relative Growth of Beaumont, Port Arthur, and Houston: 1900-1920

		1900	1910	1920	% Change 1900-1920
(1)	Beaumont	9,427	20,640	40,422	329
(2)	Port Arthur	900	7,663	22,251	2,372
(3)	Beaumont + Port Arthur	10,327	28,303	62,673	507
(4)	Houston	44,633	78,800	138,276	210
(5)	(3) as a percentage of (4)	23%	36%	45%	

Source: U.S. Census of Population, 1900-1920.

In the two decades after Spindletop, Houston itself grew rapidly, more than doubling in population. As shown in Table 2.3, however, its expansion was not unusual in comparison with that of other major Texas cities during the same period. All of those listed except Galveston, which was almost totally destroyed by a hurricane in 1900, shared relatively equally in the shift in population that made the Southwest (Texas, Oklahoma, New Mexico, and Arizona) one of the fastest growing areas in the nation from 1900 to 1920.[59] Houston's expansion is somewhat more impressive when viewed in such a national context. It grew much faster than New Orleans, the city that it would soon rival as the major Gulf Coast port. It also far outpaced most of the larger eastern and midwestern cities, a fact illustrated by its climb up in the hierarchy of cities from 85th in size in 1900 to 45th in 1920.[60]

The coastal region grew along with Houston and the Beaumont-Port Arthur area. (See Table 2.4). With the surging oil industry providing a potent impetus for growth, its percentage of the total sectional population rose from eighteen in 1900 to twenty-nine in 1920. The cotton and mixed regions in the interior grew much more slowly than did the coast. Although it had not yet attracted the overwhelming proportion of sectional population that it would later capture, the coastal region nonetheless registered impressive gains from 1900 to 1920. The coastal cities that would loom so large in future sectional growth had also been the primary beneficiary of expansion during the early period.

The competition among these cities for population and industry was, for a time, spirited. The explosive expansion of Beaumont in the boom

Table 2.3. The Growth of Houston Compared to Other Cities in Texas and Louisiana: 1900–1920

	1900	1920	% Change 1900–1920
Houston	44,633	138,276	210
Galveston	37,788	44,255	17
Dallas	42,638	158,976	272
Fort Worth	26,688	106,482	299
San Antonio	53,321	161,379	203
New Orleans	287,104	387,219	35
Texas	3,048,710	4,663,228	53
U.S.	75,994,575	106,466,000	40

Source: U.S. Census of Population, 1900, 1920.

days after Spindletop caused some observers to predict that the locus of economic activity on the Texas Gulf Coast had permanently shifted from Houston to that city. Indeed, after surveying its growth from 1901 to 1907, the Beaumont Board of Trade proclaimed its city "the future metropolis of the South." In reality, Beaumont's brief interlude as the principal focus of regional development was short-lived. The Beaumont-Port Arthur area was not to be a great future "metropolis." Rather, it was to be the site of numerous large refineries. Despite the fact that the Spindletop boom had attracted both the attention and the investment funds of the nation to these two cities in the years immediately after 1901, Houston successfully retained its role as the primary regional metropolis.

It did so for various reasons. Before 1901 Houston had a substantial lead over Beaumont in transportation, communications, and finance. Cotton, timber, and railroads had established this advantage, and all three continued to prosper. Cotton was especially dynamic. In the sixteen years before 1900, gross shipments of cotton from Houston averaged about one million bales annually. In the sixteen years after, almost 2.5 million bales were shipped yearly.[60] In addition, Houston benefited from the Spindletop boom. Many major investors, such as those involved

Table 2.4. Regional and Sectional Population Growth: 1900-1920

	1900	1910	1920	% Change 1900-1920
Coast	161,900	251,700	389,850	141
Cotton	433,476	443,770	473,790	9
Mixed	140,856	152,988	174,569	24
Timber	189,000	261,800	303,921	61
Total	924,745	1,111,061	1,342,130	45
Total -Coast	762,845	859,361	952,280	25

See pages 20 and 21 above for an explanation of these divisions.

in the Houston Oil Company, came from that city. In addition, men and equipment from the East generally came into the region through Houston, whose railroad connections were far superior to those of Beaumont. Also, after Spindletop began to decline, several important new fields opened up south and east of Houston, which became a primary center for the distribution and later for the manufacture of oil tools used in drilling new wells. As the center of producing activity moved away from the Beaumont-Port Arthur area, it was increasingly isolated as a refining area. The other branches of the vertically integrated operations of the largest oil companies gradually moved out of Beaumont. Houston's central location and its previous experience in serving the financial and administrative needs of the timber and cotton industries made it an attractive site for the main offices of regional oil companies. Joseph Cullinan and Texaco were among the first to recognize the advantages of Houston. As early as 1905, Cullinan predicted that "Houston seems to me to be the coming center of the oil business for the southwest."[61] In 1908 he helped fulfill his own prediction by moving Texaco's general offices to Houston. Cullinan also encouraged the supply company in which he had an interest, the Petroleum Iron Works Company, to locate a regional branch office there.[62] Other oil companies and supply firms followed later. Houston had no large refineries during the early period. But as the financial, commercial, and transportation capital of the region, it nonetheless shared in the profits of the growth of refining, just as it had done previously with the timber industry.

The outlines of a new regional economic pattern were thus clear, with the Beaumont-Port Arthur area as a major refining center, Port Arthur as a nationally important oil port, and Houston as the administrative and financial center for most of the oil companies active in the region. Crude oil production was an important part of this system, but pipelines from the major refineries reached far outside the region to obtain much of the crude refined in the coastal plants. These pipelines and a growing fleet of tankers that carried petroleum products from the Gulf Coast to the East Coast had gradually replaced the railroads as the prime transporters of oil, but the railroads continued to play a substantial role in the regional economy. Cotton and timber also remained important. Such pre-oil industries, however, no longer held the future of the coastal region. The oil industry had surpassed them as the central economic force affecting regional development. Fifteen years after Spindletop, the large refineries remained concentrated in the Beaumont-Port Arthur area, but already by 1916 the coastal region had emerged as a significant center of national refining. In the next twenty-five years, this nucleus of refineries would grow and spread throughout most of the coastal region, making it the largest refining center in the world and, in so doing, intensifying the process of oil-led regional development that had begun in the early years after Spindletop.

NOTES

1. Harold Williamson, et al., *The American Petroleum Industry: The Age of Energy 1899–1959* (Evanston, Illinois, 1963), p. 7. See also, Ralph and Muriel Hidy, *Pioneering in Big Business, 1882–1911* (New York, 1955).

2. Ralph Andreano, "The Emergence of New Competiton in the American Petroleum Industry Before 1911," unpublished doctoral dissertation, Northwestern University, 1960.

3. Henrietta Larson and Kenneth Porter, *History of Humble Oil and Refining Company, A Study in Industrial Growth* (New York, 1959), p. 13. Also, Harold Williamson, *The Age of Energy,* p. 16.

4. Ralph Andreano, "The Emergence of New Competition," p. 97, and Harold Williamson, *The Age of Energy,* p. 74.

5. There is extensive correspondence in both the Cullinan papers and the Texaco Archives regarding a projected merger between Gulf and Texaco from 1904 to 1906. For negotiations between Gulf and Standard, see William L. Mellon, *Judge Mellon's Sons* (Pittsburgh, 1948), p. 269. For those between Shell and Standard, see interview with C. H. Ruhl dated June 2, 1905, Drawer 396, folder 3203, part 1, Records of the Bureau of Corporations, Record Group 122, National Archives, Washington, D.C. Both economic and political pressures resulted in a similar evolution of market structure in other industries once monopolized by a single firm. See, for example, Alfred Eichner, *The Emergence of*

Oligopoly (Baltimore, 1969). John O. King, *Joseph Stephen Cullinan* (Nashville, 1970), pp. 153-183.

6. Harold Williamson, *The Age of Energy*, pp. 167-205.

7. Ibid., p. 19.

8. Carl Coke Rister, *Oil! Titan of the Southwest* (Norman, Oklahoma, 1949), p. 49.

9. Inscription on the Spindletop Monument, Beaumont, Texas.

10. Secretary of State of Texas, *Biennial Report—1898* (Austin, 1899), pp. 14-54. Secretary of State of Texas, *Biennial Report—1900* (Austin, 1900), pp. 32-61.

11. These figures are as of 1902 and could therefore be inflated by post-oil expansion. However, other sources suggest that they are accurate. See, Walton and Walton, *Directory of Texas Industries* (Austin, 1903), pp. 291-295. Copy in the Texas State Archives in Austin, Texas.

12. Annual Report of the T&NO to the Texas Railroad Commission for the year ending June 30, 1901. The report is in the files of the Texas Railroad Commission, Box 4-3/695, Texas State Archives, Austin, Texas.

13. *Beaumont Journal*, April 17, 1901. All stock values listed represent authorized capital stock, which, as discussed below, was often much more than was ever issued.

14. Everette Martin, "A History of the Spindletop Oil Field," unpublished M.A. thesis, University of Texas—Austin, August, 1934. R. T. Hill, "The Beaumont Oil Field, with Notes on Other Oil Fields of the Texas Region," *Journal of the Franklin Institute* (August-October, 1902), p. 27, places the capitalization at the end of 1902 at $200 million.

15. *Oil Investors' Journal*, September 1, 1904, pp. 1-3. This is the forerunner of the present *Oil and Gas Journal*. It originated at Beaumont in 1902.

16. *Oil Investors' Journal*, January 1, 1903, p. 11.

17. Examples are easy to site. Some, like the founder of Texas Consolidated Oil Company, with authorized capital of $2 million, simply advertised valuable holdings, listed a nonexistent office in New York, sold stock and then vanished. A more interesting case was that of the Reverend J. B. Cranfill, a Baptist preacher who advertised stock in his company in Baptist magazines throughout the South. According to the *Oil Investors' Journal*, Cranfill's scheme collapsed after he was involved in a widely publicized incident which occurred on a train trip to a Baptist convention in Atlanta. When another preacher kidded him about his oil company, the Reverend Cranfill followed the man into the train's restroom and shot him. Charged with assault, Cranfill dropped out of sight, taking his oil company with him.

18. For the early history of Gulf see John McLean and Robert Haigh, *The Growth of Integrated Oil Companies* (Boston, Mass., 1954), pp. 72-80. Craig Thompson, *Since Spindletop: A Human Story of Gulf's First Half-Century* (Pittsburgh, 1951). William Mellon, *Judge Mellon's Sons*, Chapters 26 and 27.

19. These charters are on microfilm at the Texas Secretary of State's office in Austin, Texas.

20. William Mellon, *Judge Mellon's Sons*, pp. 264-265.

21. Box 2R252, "Texas Company" file, pp. 3-5, Papers of Louis Wiltz Kemp, Barker Library, University of Texas—Austin, Texas. See, also, *Texas Company Archives*, volume 13, pp. 173-74 in the Texas Company Archives, White Plains, New York, for a list of stockholders as of April 24, 1903. For the early history of Texaco, see Marquis James, *The Texaco Story, 1902-1952* (The Texas Company, 1953) and John O. King, *Joseph Stephen Cullinan*.

22. *Beaumont Enterprise*, January 31, 1903.

23. Patrick Calhoun was born at the John C. Calhoun plantation at Ft. Hill, South

Carolina in 1856. Raised in Georgia, he rose to prominence in the 1890s (with the aid of J. P. Morgan) through local railroads. In 1895 he opened an office in New York and became "a foremost corporation attorney." After several spectacular legal controversies in Texas (ownership of oil company) and in California (trial for bribery), he died in a car wreck in Southern California in 1943.

24. John O. King, *The Early History of the Houston Oil Company, 1901-1908* (Houston, 1959). Charter of Houston Oil Company and Kirby Lumber Company, dated July 3, 1901 on microfilm in Secretary of State of Texas' office, Austin, Texas. Papers of C. A. Warner, Gulf Coast Historical Association, University of Houston, Houston, Texas. See especially Box 62, Box 192.

25. Interview with F. W. Well, Drawer 352, folder 2881, part 2, Records of the Bureau of Corporations, Record Group 122, National Archives, Washington, D.C.

26. *Oil Investors' Journal*, January 1, 1903, pp. 2-3.

27. Oil Exchange and Board of Trade, Beaumont, Texas, *The Advantages and Conditions of Beaumont and Port Arthur Today* (Beaumont, 1902) in Tyrrell Historical Library, Beaumont, Texas.

28. The source for these biographical sketches is two looseleaf notebooks entitled *Biography* in the Texas Company Archives, White Plains, New York.

29. Joseph Cullinan (Beaumont) to H. L. Scrafford (Pittsburgh) letters dated June 5, 1906 and June 18, 1906 in Drawer A-2, Papers of Joseph Stephen Cullinan, Gulf Coast Historical Association Archives, Houston Metropolitan Research Center, Houston Public Library.

30. William Mellon, *Judge Mellon's Sons*, pp. 161, 275.

31. Mobil Oil Company Publications Staff, *History of the Refining Department of the Magnolia Petroleum Company* (Beaumont, Texas, n.d.), pp. 21-24. Little is known about the builder of the refinery, George Burt, except that he constructed eastern refineries and worked on the unsuccessful French efforts at building a Panama Canal. But his close connections with Standard are obvious in his subsequent actions.

32. Cullinan to Schlaet, telegram dated October 27, 1904, *Texas Company Archives*, vol. 36, p. 55.

33. *Biography*, Texas Company Archives.

34. Interview with Will R. Wilson, Sr. in September, 1952, Transcript number 78, Oral History of the Texas Oil Industry, Barker Library, University of Texas—Austin, Texas. See also James Clark and Michel Halbouty, *Spindletop* (New York, 1952), p. 100.

35. William Mellon, *Judge Mellon's Sons*, p. 276. See also, Robert Henriques, *Marcus Samuel* (London, 1960), pp. 269-270, 462-467. For Lone Star and Crescent, see *Oil Investors' Journal*, January 1, 1903, pp. 2-3.

36. Robert Henriques, *Marcus Samuel*, pp. 340-361.

37. The Gulf statistics are from: Interview with C. H. Markham, May 8, 1905, Drawer 252, folder 2881, part 1, Petroleum Investigation, Bureau of Corporations, Record Group 122 in the National Archives. For Standard refining see Ralph and Muriel Hidy, *Pioneering in Big Business 1882-1911*, pp. 414-415.

38. *Oil Investors' Journal, The Oil Investors' Journal Dope Book* (Tulsa, 1907), p. 13. Copy in Mobil Oil Corporation Archives, Mobil Oil Building, New York City. For a concise history of Gulf's early years, see John McLean and Robert Haigh, *The Growth of Integrated Oil Companies*, pp. 72-80, 682.

39. These capacities are taken from a combination of two sources that listed the capacities of individual refineries throughout the nation as of 1916: H. G. James, *The Refining Industry of the United States* (Derrick Publishing Company: Oil City, Pennsylvania,

1916), pp. 6-11; P. L. McGreal, "Oil Refineries in Texas and Louisiana," memo dated February 15, 1917, *Texas Company Archives,* volume 51, pp. 21-28.

40. Schlaet (New York) to Cullinan (Beaumont), letter dated January 1, 1903 in Box 2R252, Kemp Papers.

41. Interview with Cullinan on May 8, 1905, Drawer 397, File 3208, Bureau of Corporation's Petroleum Investigation. The figures given for Texaco sales from May 1, 1904-May 1, 1905 are the following:

Standard Oil, New York	1,350,682 barrels
Texas Fuel Oil Co., Ltd., New Orleans	759,813
Gulf Colorado and Santa Fe R.R.	849,369
Rio Bravo Oil Company	980,792

42. Schlaet to Cullinan, letter dated June 7, 1902 in Bound Letter Press of Arnold Schlaet, Texas Company Archives. Standard's main offices were at 26 Broadway in New York City.

43. Schlaet to Cullinan, letter dated September 29, 1903, *Texas Company Archives,* volume 19, pp. 12-14.

44. Cullinan to Schlaet, letter dated August 2, 1904, *Texas Company Archives,* volume 33, p. 86. Also, Schlaet to Cullinan, letter dated December 24, 1903, *Texas Company Archives,* volume 21, p. 157.

45. Cullinan to Schlaet, letter dated May 4, 1904, *Texas Company Archives,* volume 30, p. 54.

46. For refinery size see Appendix 1. For information about the pipeline, see *Oil Investors' Journal, Dope Book,* p. 13.

47. "The Port Arthur Works," submittals by Texaco, the Temporary National Economic Committee (1937-1940), Texas Company Archives.

48. Interview with F. W. Weller, May 8, 1905, Drawer 352, folder 2881, part 2, Bureau of Corporation's Petroleum Investigation. This interview gives a list of those who bought oil from Security.

49. These incidents are discussed in detail in Mobil Oil, *History of the Refining Department of the Magnolia Petroleum Company.*

50. H. G. James, *The Refining Industry of the United States,* p. 8.

51. For information on the General refinery, see *Miscellaneous Sketches,* volume 2, pp. 206-209, in Texas Company Archives. For the Forward Reduction Company, see *Oil Investors' Journal,* August 29, 1902.

52. U.S. Census of Manufactures, 1900 and 1915.

53. American Petroleum Institute, *Petroleum Facts and Figures, 1971 Edition* (Washington, 1971), p. 154.

54. By 1905 railroads in the Port Arthur-area operated almost 850 tank cars, with the T&NO owning well over half of these. See, Annual Report of the T&NO to the Texas Railroad Commission, 1901 and 1905.

55. Marilyn Trevey, "The Social and Economic Impact of the Spindletop Boom in Beaumont in 1901," unpublished M.A. thesis, Lamar State College, 1974, Beaumont, Texas, estimates that one million dollars in new construction occurred in Beaumont between Spindletop's discovery and August, 1901. See also, *Beaumont: The City Awake* (Beaumont Chamber of Commerce, 1906) and *Advantages and Conditions of Beaumont and Port Arthur Today* (Beaumont Oil Exchange and Board of Trade, 1902). Both are in the Tyrrell Historical Library, Beaumont, Texas.

56. Hamilton Easton, "The History of the Texas Lumbering Industry," unpublished doctoral dissertation, University of Texas—Austin, 1947, p. 82. See also, W. W. Dexter, *The Coast Country of Texas* (Southern Pacific Passenger Department, 1903), pp. 36–40, in State Archives, Austin, Texas.

57. *Port Arthur Herald,* May 4, 1901, p. 1; Jones Advertising Company, *Souvenir of Beaumont, Texas* (Dallas, 1903), p. 9, in Tyrrell Historical Library; *Advantages and Conditions of Beaumont and Port Arthur Today; East-Southeast Texas on the Line of the Texas and New Orleans Railroad* (T&NO, 1907), in Barker Collection; "Argument of the Beaumont Oil Exchange and Board of Trade Recommending Proposed Improvement of Sabine Line Ship Channel," pamphlet dated November, 1902, pp. 22-29, in Tyrrell Historical Library.

58. These comparisons are taken from two pamphlets, *Advantages and Conditions of Beaumont and Port Arthur* (1902) and *Beaumont: The City Awake* (1906). (See footnote 57.)

59. Harvey Perloff et al., *Regions, Resources, and Economic Growth* (Baltimore, 1960), p. 13.

60. Houston Cotton Exchange and Board of Trade, *Fifty Years a Cotton Market: 1874–1924* (Houston, 1924), p. 50.

61. Cullinan (Beaumont) to Schlaet (New York), letter dated February 3, 1905, Drawer 1-A, Cullinan Papers.

62. Cullinan to C. H. Todd (Petroleum Iron Works), letter dated February 23, 1905, Drawer 1-A, Cullinan Papers.

Chapter III

The Middle Years (1916–1941): The Heyday of Regional Refining

World War I introduced a period of extraordinary expansion in the oil industry. With its close ties to this dynamic industry, the Gulf Coast refining region prospered throughout the war years, the 1920s, and even—at least in comparison to the remainder of the nation—the depression years of the 1930s. During this middle period, the region's original refining complex in the Port Arthur-Beaumont area continued to grow, but the most significant new development was the emergence of an equally large cluster of refineries in the Houston area. Accompanying the sustained growth and geographical spread of refining were technological changes in the refining process that greatly increased both the value of goods produced in the region and the level of investment required to produce them. As a result, in its heyday from 1916 to 1941, refining played a more important role in the regional economy than it had played in the earlier period or would play after World War II. Its rapid growth during those middle years encouraged significant changes in the regional and sectional distribution of population and of industrial activities while preparing the way for more far-reaching changes in the post–World War II era.

THE GROWTH AND SPREAD OF NATIONAL REFINING

Throughout the twentieth century the demand for refined petroleum products has steadily expanded. Amid this general long-run trend, however, the years from 1916 to 1941 stand out as an era of exceptional growth. The rising consumption of petroleum products in this period—

and especially in the 1920s—reflected an increase in the use of all forms of energy.[1] The most spectacular gains were registered in the sale of gasoline and fuel oils. Neither was a new product but both enjoyed tremendous expansion from 1916 to 1929. The number of automobiles in the nation had expanded gradually in the first decades of the century, but the biggest surge came after World War I. In 1916 there were about 3.6 million motor vehicles registered in the United States; thirteen years later, in 1929, there were 26.7 million.[2] Over this period, the domestic demand for motor fuel rose by a factor of five. In addition, the demand for fuel oils more than tripled during the same time. Of course, the Great Depression dampened the nation's appetite for refined goods, and not until 1935 did the total demand for petroleum products reach the level attained in 1929. But another surge in demand in the late 1930s followed these lean years, providing a fitting climax to an era of very rapid expansion. (See Table 3.1.)

The petroleum industry met part of this rising demand for refined petroleum products in the 1920s and the late 1930s by introducing far-reaching technological changes that increased the amount of gasoline obtained from each barrel of crude oil distilled. (See Table 3.2.) Indeed, the growing demand for gasoline encouraged several important breakthroughs in refining technology, as refiners sought new ways to "crack" hydrocarbon molecules in ways that would produce more and better gasoline.[3] The first major innovation, the Burton thermal cracking process pioneered by Standard of Indiana in 1913, opened a twenty-year period of improvement in the thermal cracking process. The pace of change in these years led one expert on refining to comment in 1934 that "more technological changes in refining took place in the last fifteen years than over its entire previous history."[4] Research under way as he spoke led to a second major innovation in refining technique, catalytic cracking, that introduced a period of even greater technological change. These new methods of refining played an important role in meeting the growing demand for gasoline. In 1918, cracking units produced less than 15 percent of all gasoline refined in the United States. This figure had grown to almost 40 percent in 1929 and to more than 55 percent in 1941.[5]

Technological changes alone, however, could not supply the expanding market for refined petroleum products. Widespread construction of additional refining capacity was also required. In 1918, the nation's refineries could process about 1.3 million barrels of oil per day. By 1941,

Table 3.1. U.S. Demand For Petroleum Products: 1919–1941

Year	Thousands of Barrels Per Day	Percentage Change
1919	1026	----
1920	1245	17.6
1921	1254	.7
1922	1455	13.8
1923	1787	18.6
1924	1878	4.8
1925	1992	5.7
1926	2139	6.9
1927	2201	2.8
1928	2350	6.3
1929	2577	8.8
1930	2539	-1.5
1931	2473	-2.7
1932	2282	-8.4
1933	2379	4.1
1934	2520	5.6
1935	2694	6.5
1936	2985	9.7
1937	3203	6.8
1938	3114	-2.9
1939	3371	7.6
1940	3625	7.0
1941	4071	11.0

Source: American Petroleum Institute, *Petroleum Facts and Figures*, 1971 edition (Washington, 1971), pp. 285–287.

Table 3.2. Average Yield of Petroleum Products Per Barrel of Crude Oil Refined in the United States (Percent of Each Product Per Barrel of Crude Refined)

Year	Gasoline	Kerosene	Fuel Oil[a]	Residual[a]	Lubricating	Others
1899	12.9	57.6	14.0	--	9.1	6.5
1904	10.3	48.3	12.8	--	11.6	17.0
1909	10.7	33.0	33.6	--	10.7	12.0
1914	18.2	24.1	46.5	--	6.6	4.6
1918	25.3	13.3	53.5	--	6.1	1.8
1921	27.1	10.4	51.9	--	4.7	5.9
1926	34.4	7.8	46.9	--	4.1	7.5
1931	44.1	4.7	9.4	28.2	3.0	10.6
1936	43.8	5.2	11.7	26.7	2.9	9.7
1941	44.2	5.2	13.4	24.2	2.8	10.2

Source: American Petroleum Institute, *Petroleum Facts and Figures*, 1971 edition (Washington, 1971), p. 203.

[a] Until 1931, "fuel oil" and "residuals" were reported together.

more than 3 million more barrels per day could be refined. (See Table 3.3.) The amount of crude oil annually refined in these plants more than quadrupled from 1916 to 1929 in a spectacular surge of expansion that has not been approached since in a similar period of time. Over the entire middle period (1916 to 1941), annual crude runs in refineries in the United States rose by approximately 470 percent.[6]

Such aggregate statistics are suggestive, but other measures of expansion illustrate more vividly the magnitude of the growth of refining between the wars. In the oil industry, one obvious point of reference is the Standard Oil Company. "The Octopus," which remains to the present a potent symbol of monopoly power and industrial giantism, dominated all phases of the industry for most of the years before 1916. In 1911, the year of its dissolution and of its greatest output, the total

Table 3.3. The Growth of U.S. Refining Capacity 1918–1941

Year	Operating Capacity in Barrels Per Day	Percentage Increase
1918	1,295,115	—
1921	1,854,590	43
1926	2,834,292	53
1931	3,624,992	28
1936	3,966,616	9
1941	4,496,843	13

Source: American Petroleum Institute, *Petroleum Facts and Figures,* 1971 edition (Washington, 1971), p. 140.

refining capacity of all refineries operated by Standard was approximately 300,000 barrels per day. In 1929, the refineries in Jefferson County alone surpassed this output. In 1936, those in Harris County (Houston) also exceeded 300,000 barrels per day. By 1941—only thirty years after the dissolution of Standard—refineries on the upper Texas Gulf Coast alone more than tripled its greatest daily capacity.[7]

This last comparison suggests another significant change that accompanied the expansion of refining capacity in this period: the increasing importance of the Gulf Coast. Indeed, as shown in Table 3.4, in the 1920s and the 1930s this area became the largest refining center in the nation. Its share of the total national capacity expanded steadily until the 1930s, when a burst of refinery construction pushed its percentage past that of the West Coast and far beyond that of the traditional refining areas in the East. In the years between 1916 and 1941, and especially in the late 1930s, this rapid ascent of Gulf Coast refining stands out as the major change in the location of the nation's refining capacity.

The primary reasons for this shift after 1916 were similar to those that originally prompted Gulf, Texaco, and Standard of New Jersey to build refineries near the Spindletop field. For such large companies, microeconomic decisions about refinery location were made in the context

Table 3.4. Changing Geographical Location of Operating Refining Capacity in the United States, 1916–1941 (Percentage of National Capacity in Each Region)

Year	Region								
	1	2	3	4	5	6	7	8	9
1916	13.8	18.5	7.0	10.7	14.9	6.0	.6	3.8	20.3
1918	17.2	19.7	5.8	6.8	18.7	4.4	.5	2.8	23.9
1921	15.0	21.8	4.4	8.6	14.3	3.4	1.7	3.9	17.4
1926	18.0	16.4	3.7	8.9	12.4	5.1	3.2	4.6	26.1
1931	18.1	16.6	3.7	11.3	13.0	7.2	6.5	4.1	23.5
1936	20.4	16.5	4.1	12.3	11.9	8.7	2.4	2.3	21.3
1941	27.9	15.4	3.6	15.9	9.0	5.3	1.9	2.6	18.3

Source: The chart was constructed by taking the operating capacity reported for each region and dividing it by the operating capacity reported for the entire nation. These capacities are taken from the yearly bulletin of refining statistics published by the Bureau of Mines. Because of difficulties caused by the lack of comparable data, the years 1916–1926 for regions 2–5 were computed using total crude consumed, not operating capacity. Figures for crude consumed were taken from *Petroleum Facts and Figures,* 1959 edition (Washington, 1959), p. 106. Citations for specific Bureau of Mines *Bulletins* can be found in the bibliography.

These are the standard Bureau of Mines refinery districts, with minor alterations due to the changes which the Bureau has made over the years. For example, New Mexico has been reported separately from the Rocky Mountain region since 1941. See *Petroleum Facts and Figures* for 1959, p. 92, for a map and additional information as of that year.

1. Gulf Coast of Texas and Louisiana.
2. East Coast: contains all Atlantic Coast states and the eastern portions of New York and Pennsylvania, including New York City and Philadelphia.
3. Appalachian: Western Virginia, the remainder of New York and Pennsylvania, and the eastern portion of Ohio, including the Cleveland area.
4. Indiana, Illinois, and Kentucky, the remainder of Ohio, Tennessee, Michigan, Minnesota, Wisconsin, North and South Dakota.
5. Oklahoma, Kansas, Missouri, Nebraska, Iowa.
6. Texas inland—all of Texas minus the Gulf coastal region.
7. Arkansas, Louisiana inland. Does not include Louisiana Gulf Coast.
8. Rocky Mountains: Montana, Idaho, Wyoming, Utah, Colorado, and New Mexico.
9. West Coast, primarily California.

of the overall needs of their vertically integrated operations. Such factors as labor and fuel costs influenced locational decisions, but each company also chose refinery locations according to the availability of crude, relatively inexpensive means of transport, and access to markets. For these integrated companies planning to sell in the large urban-industrial markets of the northeastern United States—that is, for most major American oil companies—the balance of these factors made the Gulf Coast a particularly attractive location for large refineries.

Such facilities required great quantities of crude oil. The refineries around Beaumont and Port Arthur were originally constructed near Spindletop to take advantage of the relatively plentiful supply of crude from what was, by the standards of the time, a very productive field. The decline of the Spindletop field would have meant the end of the refining complex if alternative sources of crude had not been discovered. But the opening of a succession of large southwestern oil fields became one of the primary attractions of the Gulf Coast as a refinery location.[8] Secure, predictable sources of crude from Texas, Oklahoma, Arkansas, and Louisiana meant that the large coastal refineries faced little uncertainty, at least in the middle period, in obtaining the substantial daily supplies of oil necessary to ensure continous operations.

In converting this raw material into finished products, the refineries also required substantial quantities of oil and natural gas for fuel, and the availability of these relatively inexpensive fuels on the Gulf Coast further encouraged the growth of refining there. Statistics on fuel consumption in the refineries before the 1920s are sketchy, but perhaps as much as 10 percent of the crude delivered to the refineries was used as fuel to provide the intense heat required for distillation. Bureau of Mines studies in 1927–1929 suggest that in those years the refineries on the Texas Gulf Coast consumed from 6 to 7 percent of the total oil that they processed.[9] The resulting need for reliable sources of fuel was thus an important consideration in choosing a location for a large refinery, and abundant, accessible supplies of oil and natural gas helped induce many companies to choose a Gulf Coast site.

Advantages in transportation costs for both crude and refined products gave the Gulf Coast an added locational superiority over other refining regions. Pipelines over relatively flat lands in Texas, Louisiana, and Oklahoma carried crude to the coastal region, which was only a tanker voyage away from the large markets on the eastern seaboard. Until the construction of cross-country pipelines during and after World War II, tankers were the only economically and technologically viable means for the long-haul shipment of petroleum products. The Gulf Coast was thus closer to the East in terms of transportation costs than were other major oil producing and refining areas.[10] In the interwar years, this easy access to the eastern seaboard was crucial to the Gulf Coast's expansion, since shipments to the urban Northeast absorbed up to 70 percent of its refined output. (See Table 3.5.)

Table 3.5. Shipment of Petroleum Products From the Gulf Coast to the East Coast, 1926–1941 (In Thousands of Barrels)

Year	Gasoline Shipments	Percent of Gulf Coast Production of Gasoline	Fuel Oil Shipments[a]	Percent of Gulf Coast Production of Fuel Oil	Crude Oil Shipments	Shipments of All Petroleum Products
1926	33,443	51	18,301	27	78,992	not available
1930	34,306	36	20,423	30	111,501	not available
1936	90,558	74	80,431	72	153,026	342,914
1941	130,534	64	118,543	71	147,288	430,769

Sources: For 1926 and 1930, see U.S. Bureau of Mines, *Bulletin 367* (Washington, 1932), p. 79. For 1936 and 1941, see American Petroleum Institute, *Petroleum Facts and Figures* (Washington, 1971), pp. 161 and 262.

[a] This includes "Gas Oil," "Distillate Fuel Oil," and "Residual Oil."

Derivation: The figures for "Percent of Gulf Coast Production" were obtained by dividing the Gulf Coast shipments to the East Coast (p. 262 in the 1971 edition of *Petroleum Facts and Figures*) by the total production listed in the same source under the "Texas Gulf Coast" and "Louisiana Gulf Coast" refining districts (p. 161).

The opinions of executives from two large oil companies that invested substantial sums in coastal refineries provide a good summary of the locational advantages of the upper Texas Gulf Coast. In 1927, Standard of New Jersey's Manufacturing Department compared the cost of refining at Humble's Baytown plant with that of Jersey's refineries in the New York City area. It found that Baytown had "strong advantages" in the cost of fuel, water, land, taxes, and labor; it also was well located for receiving Texas crudes and for bulk export shipment of products.[11] In 1939, the President of Pan American, testifying before the Temporary National Economic Committee, stressed the Gulf Coast's advantages over the East Coast in fuel costs, in ocean transportation for gasoline, and in flexibility of marketing. He concluded that "the recent trend among both large and small refiners has been rather strongly toward a Gulf Coast location for new refinery capacity designed for eastern markets."[12]

A combination of factors thus encouraged the selection of Gulf Coast sites for many of the plants needed to supply refined petroleum products during a period of extraordinary expansion in the demand for these goods. The resulting growth in regional refining capacity had a cumulative impact. The construction of large refineries in the region encouraged the development of specialized skills and facilities, which then helped convince still other oil companies to build their plants there. As a result of a variety of natural and economic locational advantages, the upper Texas Gulf Coast thus emerged as the largest refining center in the world.

GROWTH AND SPREAD OF REGIONAL REFINING

Much of this growth occurred outside the early center of Gulf Coast refining, the Port Arthur area, as refineries spread geographically eastward and westward. Thus, by 1941, "Gulf Coast" refining was no longer located almost exclusively in Jefferson County. Instead, by that time clusters of refineries dotted the coast from New Orleans to Corpus Christi, some 200 miles south of Houston (see Map 3.1). The focus of Gulf Coast refining nonetheless remained the upper Texas Gulf Coast. Within that region, however, Houston now challenged the Port Arthur-Beaumont area as the capital of regional refining (see Table 3.6).

This was not because the refining capacity of the latter area ceased to expand. Indeed, Jefferson County's capacity more than tripled between

Map 3.1. Refining Centers on the Texas-Louisiana Gulf Coast

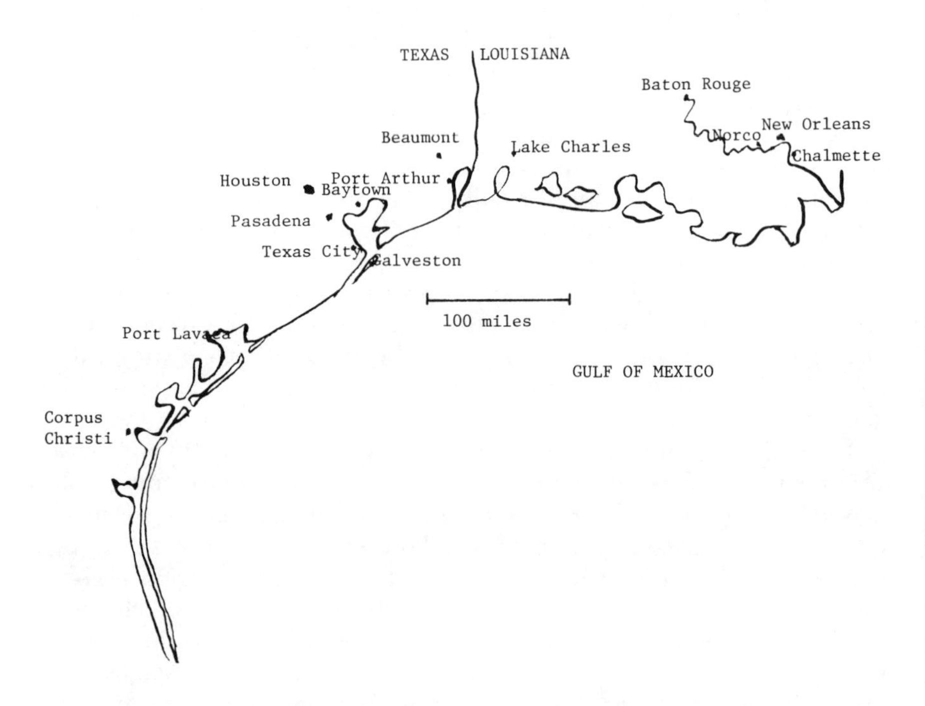

1916 and 1941. The three large refineries built between 1901 and 1905 remained the dominant plants. Gulf's Port Arthur refinery grew steadily, and it remained the largest American refinery of its parent company and one of the largest operated by any company anywhere in the world. Texaco's Port Arthur works, which had been much smaller than the neighboring Gulf Oil plant in the early period, doubled in capacity in the 1930s, also entering the ranks of the nation's largest refineries. Mobil's Beaumont refinery experienced similar expansion in the 1930s. In 1940, each of the three large Jefferson County plants had a capacity in excess of 100,000 barrels per day, a figure surpassed by only two other refineries outside of the Houston-Port Arthur region.[13] Their steady growth, the increased production of Texaco's second plant at nearby Port Neches, and the construction of several smaller refineries, by Pure Oil in 1924 and Atlantic Refining Company in 1938, meant that throughout the interwar years, Jefferson County continued to supply more than 10 percent of the entire national output of refined goods.

Table 3.6. Operating Crude Capacity of Refineries on the Texas-Louisiana Gulf Coast, 1904–1941 (42-Gallon Barrels Per Day)

Year	Total	Percent Increase	Beaumont-Port Arthur	Percent of Gulf Coast Production	Houston	Percent of Gulf Coast Production	Corpus Christi	Percent of Gulf Coast Production	Louisiana Gulf Coast	Percent of Gulf Coast Production
1904	14,400	-	14,000	98.0	200	1.0	-	-	200	1.0
1909	38,000	165.0	32,000	84.0	1,000	3.0	-	-	5,000	13.0
1914	91,000	140.0	68,000	76.0	3,000	3.0	-	-	20,000	22.0
1918	204,200	124.0	138,500	67.8	4,200	2.1	-	-	61,500	30.1
1921	269,700	32.1	148,000	54.9	33,600	12.5	-	-	88,100	32.7
1926	452,300	67.7	227,000	50.2	102,800	22.7	-	-	122,500	27.1
1931	670,500	48.2	305,000	45.5	120,500	31.4	15,000	2.2	140,000	20.9
1936	774,300	15.5	309,000	39.9	305,500	39.5	23,800	3.1	136,000	17.6
1941	1,167,400	50.8	450,600	38.6	472,800	40.5	106,500	9.1	137,500	11.8

Source: Before 1918, see Appendix 1. After 1918, yearly Petroleum Refining Information Circulars from the Bureau of Mines were used. For specific citations, see the bibliography.

Note: Texas-Louisiana Gulf Coast includes the standard Bureau of Mines refining districts. This means that the Louisiana Gulf total after 1931 includes small refineries in Mobile, Alabama.

Beaumont-Port Arthur consists of a one-county (Jefferson) area. Houston includes the Houston metropolitan area. Corpus Christi encompasses all of the Texas Gulf Coast south of the Houston metropolitan area.

Despite this impressive record, the Port Arthur area lost its earlier position of dominance in coastal refining, as the capacities of other areas on the Gulf Coast expanded even more rapidly. This geographical shift had begun in 1911, when Standard's choice of a Baton Rouge site for a major refinery and the subsequent construction of several large refineries near New Orleans seemed to signal a substantial movement to the east in the focus of coastal refining. Indeed, during and immediately after World War I, the availability of plentiful imported oil from Mexico to supply these plants prompted the *Oil and Gas Journal* to proclaim: "New Orleans to Be Refining Center on Extensive Scale."[14] As the Mexican oil fields boomed, two of the largest United States companies involved in Mexican production chose locations near New Orleans for large plants in 1916: Freeport and Tampico Fuel Oil at Meraux, Louisiana, and the Mexican Petroleum Company at nearby Destrehan. Previously constructed refineries also converted in part to the use of Mexican crude, briefly making it an important source of oil for the Gulf Coast refining complex (see Table 3.7). Because of growing conflicts between the Mexican government and the U.S. and British companies in Mexico, however, imports of oil from that country had slowed to a trickle by 1927. Both New Orleans area plants built to process Mexican oil were subsequently absorbed by major oil companies—the Meraux plant by Sinclair and the Destrehan plant by Pan American (Standard of Indiana). Neither refinery experienced the sustained growth that characterized its counterparts in Port Arthur. The New Orleans area attracted one more major refinery in 1930, Shell's plant at Norco, Louisiana. But the 1914 prophecy of the *Oil and Gas Journal* went largely unfulfilled in this period, and Standard's Baton Rouge works remained the only large refinery in Louisiana until World War II.

Had the journal replaced "New Orleans" with "Houston," its judgment would have been vindicated. The spread of refining in the interwar years was to the west, not to the east, and the Houston area was far and away the major beneficiary of this shift. The construction of large refineries in this area began in earnest several years after the completion of a ship channel connecting Houston to the Gulf of Mexico. Although opened in late 1914, the channel did not begin to attract refineries until the end of World War I. Then, in the 1920s, large plants sprang up along both sides of Houston's corridor to the sea. (See Map 3.2.)

Leading the way was the giant Baytown refinery of the Humble Oil and Refining Company. When organized in 1917, this company consolidated several local concerns then active in the booming oil fields around

Table 3.7. Foreign Crude Used in Gulf Coast Refineries, 1918–1930 (In Thousands of 42-Gallon Barrels)

Year	Refineries on the Texas Gulf Coast	Percent of Total	Refineries on the Louisiana Gulf Coast	Percent of Total	Refineries on the Texas-Louisiana Gulf Coast	Percent of Total
1918	-	-	-	-	653	1.0
1919	7,001	14.2	6,693	34.2	13,694	19.9
1920	11,199	18.3	8,978	42.1	20,177	24.5
1921	17,975	26.7	12,679	49.9	30,654	33.0
1922	22,063	28.3	9,619	35.4	31,682	30.1
1923	12,626	15.5	7,690	24.0	20,316	17.9
1924	11,053	11.8	9,002	24.5	20,005	15.3
1925	7,577	7.0	9,995	19.7	17,572	11.0
1926	3,687	3.1	11,983	23.3	15,670	9.1
1927	814	.6	7,245	14.4	8,059	4.5
1928	1,246	.9	7,552	14.3	8,798	4.6
1929	2,873	1.9	6,397	12.6	9,270	4.5
1930	1,050	.7	2,749	6.7	3,799	1.9

Sources: U.S. Department of the Interior, Bureau of Mines, "Petroleum Refinery Statistics, 1916–1925," Bulletin 280 (Washington, 1927), pp. 16–17. See, also, Bulletins 289 (1926–7), 318 (1928), 339 (1929), 367 (1930).

Houston. Two years later, however, Humble became affiliated with Standard Oil Company (New Jersey) and rapidly moved into the ranks of the major oil companies. During the 1920s, a series of expansion programs increased the capacity of Humble's Baytown plant from only 10,000 barrels per day in 1921 to 125,000 in 1931.[15] By that date it had become the largest refinery on the Gulf Coast, a distinction it held for several decades.

Two other large refineries—constructed by Sinclair Oil Company in

Map 3.2. Refineries in the Houston Area, 1918–1926

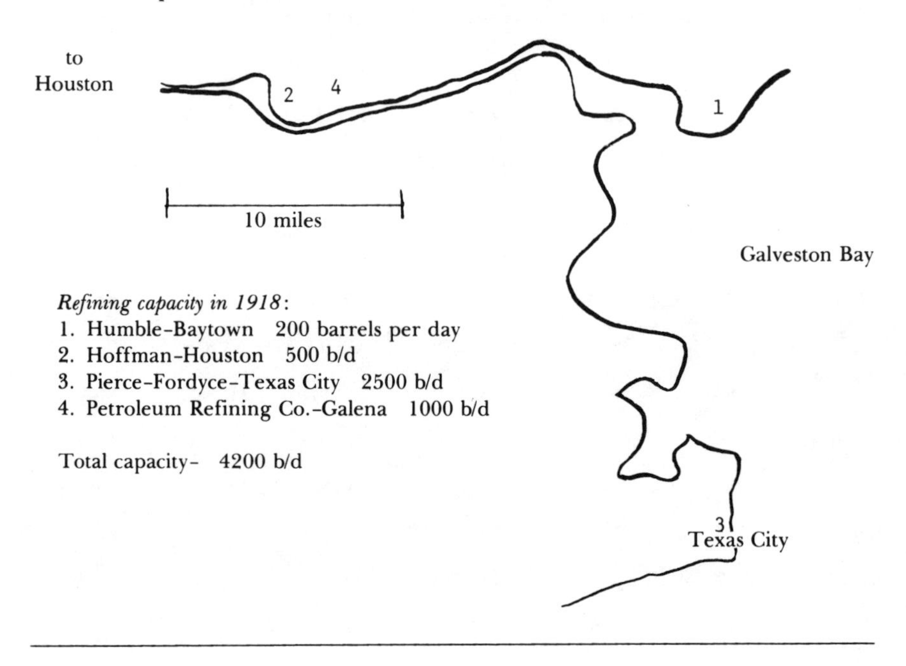

Refining capacity in 1926:

1. Humble-Baytown 50,000 b/d
2. Galena-Signal-Galena 10,000 b/d
3. Sinclair-Sinco 16,000 b/d
4. Freeport-Bryanmound 9,000 b/d
5. Deepwater-Houston 2,500 b/d
6. Keen and Woolf-Clinton 1,500 b/d
7. Crown Central-Pasadena 3,000 b/d
8. Deepwater-Laporte 800 b/d
9. Terminal-Texas City 10,000 b/d

Total capacity- 102,800 b/d

Houston in 1920 and Crown-Central in Pasadena in 1924—and numerous smaller refineries grew along either side of the ship channel in the 1920s. The 1930s, however, brought even more rapid expansion in the Houston area. Fed by the construction of large plants by Shell (Deer Creek, 1929), Pan American (Texas City, 1934), Republic (Texas City, 1932), and Eastern States Petroleum (Houston, 1937), the Houston area's capacity increased by more than 360 percent from 1931 to 1941. As a result, in 1941 this area contained an even larger refining capacity than did the Port Arthur area.

Like Houston, Corpus Christi did not emerge as a refining center until after the completion of its ship channel in 1926. Humble was the first large company to migrate southward to the Corpus Christi area, constructing a small refinery at Ingleside in 1929. A surge of construction in the late 1930s established a cluster of plants near Corpus Christi. But in the interwar years, both its geographical isolation and the relatively small size of the refineries at Corpus Christi made it less an extension of the larger refining complex on the upper Texas Gulf Coast than a separate enclave of refineries.

As refining spread geographically in the coastal region, changes in technology amplified its impact on the regional economy. The introduction of new "cracking" techniques resulted in major alterations in the construction and operation of large refineries nationwide, but the impact of such changes were especially pronounced on the upper Texas Gulf Coast, which in the middle period contained as much as 25 percent of the national cracking capacity.[16] The effects of such changes were often ignored and became apparent only over the long run, but they were, nonetheless, striking. Cracking units were much more expensive to build and maintain than the simpler stills of earlier refineries (see Table 3.8). This meant a greatly increased level of refinery-related investments in the coastal region. Texaco's Port Arthur works alone, for example, represented an investment of almost $60 million in 1937.[17] One result of this investment was an increase in the value of the refined output, which, from a regional perspective, translated into a greater value for exports. Technological change also brought gradual changes in the type and amount of labor required to run the refineries, thus altering the makeup of the regional work force. Finally, these costly changes in the techniques of production indirectly encouraged a trend toward concentration of regional refining capacity in the hands of the firms which could best afford to implement them, the large vertically integrated oil companies.

Table 3.8. Construction Costs for the First Commercial Cracking Units Incorporating New Processes, 1913–1942

			At Time of Construction:		
Process	Year	Capacity Per Day in Barrels	Construction Costs in Dollars	Construction Index (1939 = 100)	In 1939 Dollars
Burton	1913	88.5	6,750	52.1	12,950
Tube and Tank	1922	570	90,000	82.7	108,900
Houdry	1938	6,750	2,191,000	100.0	2,191,200
Fluid	1942	12,750	2,060,000	109.0	1,889,000

Source: John Enos, *Petroleum Progress and Profits: A History of Process Innovation* (Cambridge, Mass., 1962), p. 241. Reprinted from *Petroleum Progress and Profits* by John Enos by permission of the MIT Press, Cambridge, Massachusetts.

This trend toward greater concentration of ownership was not, of course, solely the result of the imperatives of technology. Rather, it reflected a variety of organizational, competitive, technological, and even political forces that encouraged the rise of an oligopolistic market structure in oil. The pattern of ownership in the interwar years was not entirely new, since in some ways it merely extended the pattern that had arisen before World War I. The dynamic young firms of the early period—Gulf, Texaco, and Magnolia—had merely matured into major, vertically integrated oil companies. Then, in the 1920s and 1930s, four other such companies—Jersey Standard, Shell, Standard of Indiana, and Sinclair—had build large plants in the Houston area. In 1940, these seven firms were all among the eight largest refining companies in the nation. Their commitment to the coastal region is best illustrated by one statistic: as of 1940, the largest refinery owned by all of these firms except Standard of Indiana was located in the Houston-Port Arthur region; Standard of Indiana's second largest plant was there. At that time, these seven companies operated about 78 percent of the total refining capacity in the coastal region. The addition of the next five largest companies active in the region (Pure Oil, Atlantic Refining, Crown Central, Republic, and Eastern) boosts this figure to more than 93 percent.[18] A defining characteristic of regional refining during the middle period was thus its control by the largest oil companies in the nation.

By the end of the interwar period, these companies had a very substantial stake in the coastal region, where they had each built at least one major refinery. During the heyday of regional refining, they had invested great sums in the expansion of existing plants, the construction of new ones, and the introduction of expensive new refining processes. These refineries played an important role in the vertically integrated operations of the large oil companies that owned them, and, in so doing, they came to play an equally important role in the economy of the coastal region.

INDUSTRIAL AND URBAN GROWTH IN THE HEYDAY OF REFINING

Spurred by the growth of refining, the coastal economy expanded rapidly for most of the period from 1916 to 1941. Indeed, the region's pattern of development in these years closely paralleled that of its refineries. Industry grew especially rapidly during the 1920s, with the Houston Ship Channel area registering the greatest gains. Although the region did not avoid the effects of the Great Depression, it did recover earlier than most of the nation. After a brief pause in the early 1930s, a strong period of expansion closed out the decade. Throughout these years, the growth of all phases of the oil industry was the primary impetus for the region's sustained development, and petroleum refining in particular played a central role in shaping regional industrialization.

The economic preeminence of refining was most clearly evident in Jefferson County, where the expansion of existing plants and the construction of several new ones increased their already considerable economic role. The Beaumont Chamber of Commerce summarized the local role of the refineries in 1925 as follows: "To get an accurate estimate of the value of the refining business to this district would be a task next to impossible. . . . Plants in Jefferson County refine more than half the oil refined in the state of Texas. These plants regularly employ in excess of 16,000 men with an annual payroll of more than forty-six million dollars."[19] At about the same time, the Port Arthur Chamber of Commerce coined the phrase, "We Oil the World," to publicize the city's importance in the world petroleum industry. Although both Chambers must be excused for a bit of exaggeration, they could hardly overstate the importance of refining to the Beaumont-Port Arthur area. The large refineries throughly dominated the industrial activity of this eastern portion of the coastal region.

The Harris County area (Houston) to the west had a somewhat more mixed economy, but refining was the major industry there also. Houston had earlier prospered primarily as a distribution center, financing and shipping goods rather than manufacturing them. But the growth of industry along the ship channel in the 1920s introduced a new era of manufacturing in the city's development, and oil refining was the earliest and most substantial industry drawn to the ship channel. In reviewing the remarkable growth of Houston in the interwar years, an article in *Fortune* magazine succinctly summed up the importance of oil by suggesting that "without oil Houston would be just another cotton town."[20]

Two tables offer statistical support for such comments. Table 3.9 shows the dominance of petroleum products in water shipments from all the ports in the refining region in 1924 and 1933. Table 3.10 provides an estimate of the contribution of petroleum refining to the total manu-

Table 3.9. Foreign Exports and Coastwide Shipments From the Costal Region,[a] 1924 and 1933 (In Thousands of Short Tons)

	1924	Percent of Total	1933	Percent of Total
All goods	23,965	100	44,551	100
Petroleum products	18,902	79	41,001	92
Crude	n.a.	n.a.	21,153	(47)
Refined	n.a.	n.a.	19,769	(44)
Chemicals	0	0	.8	-
Sulphur	1,288	5	958	2
All grains	1,205	5	203	-
Rice	90	-	131	-
Lumber products	462	2	127	-
Cotton products	1,769	7	1,422	3

Source: Department of the Army, Corps of Engineers, *Waterborne Commerce of the United States, 1958, Part II: Gulf Coast, Mississippi River System and Antilles* (Washington, 1958).

[a]Included are the ports of Houston, Beaumont, Port Arthur, Orange, Texas City, and Galveston.

Table 3.10. Petroleum Refining as a Percentage of All Industrial Activity in Harris and Jefferson Counties in 1929

	Harris County	Jefferson County	Total
Value added by all manufacturing (in millions of dollars)	93.5	61.6	154.6
Value added by refining (in millions of dollars)	28.0[a]	49.2	77.0[a]
Refining as percentage of value added	30[a]	79	50[a]
Total manufacturing workers	26,213	16,423	42,636
Workers in refining	7,000[a]	12,158	19,000[a]
Refining as percentage of workers	27[a]	74	45[a]

Source: U.S. Census of Manufacturers, 1929.

[a] All of these figures contain an estimate for Harris County refining activity. In 1929, the *Census of Manufacturers* omitted the category "Petroleum Refining" for Harris County for fear of revealing information about individual firm's employment and output. My estimate for Harris County was derived by multiplying the Jefferson County statistics on value added by refining and employment in refining by 57 percent, which is the refining capacity of Harris County divided by that of Jefferson County (as of 1929).

facturing output of the region. Numerous gaps in the Census of Manufacturers prevent a systematic analysis of the economic impact of refining, at least until the post-World War II era. The 1929 Census of Manufacturers omits detailed statistics on refining in Harris County because "no figures can be given for the most important industry—namely 'petroleum refining'—without disclosing the data reported by individual establishments."[21] Table 3.10, which includes estimates of these missing figures as well as the actual census reckoning of the economic contributions of refining in Jefferson County, suggests the extent to which refining dominated regional economic life as of 1929. From then until 1940, the refining capacity of the Houston area quadrupled, and statistics on

1940, if available, would no doubt reflect this change by showing a dramatic increase in both the value added and the employment of the refineries. Even from estimates based on 1929 information, however, it is clear that, during its heyday, refining was far and away the largest single industry in the coastal region.

The refineries, of course, were not isolated from the remainder of the petroleum industry, and other, related branches of the industry were also significant forces in regional development. Such activities will be treated in detail in Chapter V, but several in particular should be acknowledged as of special importance during the interwar period. Then, as throughout the century, oil production in fields in or near the coastal region was very important,[22] and Houston's second largest industry in these years was the manufacture of oil field tools and machinery.[23] Oil-related construction also provided a substantial stimulus for regional growth. During this period of rapid refinery expansion, the construction of refineries became a significant industry of itself, pumping large sums of money into the regional economy. Often those workers who built or expanded a refinery were then hired to operate it, but specialized construction companies and crews also developed.[24] A final type of construction begun in this period would later become increasingly important in Houston: by 1942 there were already at least six large downtown office buildings that housed petroleum companies.[25] Within the coastal region, the large refineries were both the center of and symbol for most of these oil-related activities. The "propulsion" generated by the petroleum industry as a whole drove regional development at near breakneck speed during the heyday of refining.

OTHER INDUSTRIES

Other regional industries continued to expand along with—and at times in spite of—petroleum refining during its heyday. The most significant pre-oil industries, cotton and timber, remained important, but their growth in the interwar period—and especially during the Great Depression—did not match that of the petroleum-related sector of the regional economy.

Fortune's assertion that Houston would have been "just another cotton town" without oil was close to the truth. However, the city would at least have been a large and thriving "cotton town." Although the growing of cotton gradually moved to the northern and western portions of Texas, Houston remained an important shipping point because it was located at "the little end of the funnel that drained all of Texas and the Oklahoma territory."[26] These were the words of William Clayton, a partner in Anderson, Clayton & Co. In 1916, this company moved its headquarters to Houston, where it became the most important processor and shipper of cotton in the coastal region. Pursuing a policy of aggressive expansion in the United States and in European markets that had been disrupted by World War I, Anderson, Clayton & Co. became one of the largest cotton brokerage firms in the world in the 1920s. Its prosperity and that of other smaller firms was reflected in the growth of cotton exports from the region. By the late 1920s, the port of Houston had surged past all other ports to become the leading exporter of cotton in the nation. Its sister port on the Texas Gulf Coast, Galveston, also remained a major cotton port, challenging New Orleans as the second largest exporter of cotton.[27] Because cotton required little processing before its shipment, these large exports did not generate substantial industrial activity. But significant economic benefits nonetheless accrued to the coastal region from its role as a center for cotton finance and shipping.

Timber also continued to exert a considerable influence on the regional economy. While prospering during the heyday of refining, however, the region's timber industry also underwent significant changes. The Sabine-Neches area remained the center of the region's timber-related activity, but Orange, Texas, a small city north of Port Arthur on the Sabine River, reemerged as a processing and shipping point for lumber. This shift was partly the result of Port Arthur's severe oil pollution, which made that port unfit for timber by ruining logs that were floated into the area.[28] Timber exports declined as the burgeoning demand for timber within the rapidly growing region absorbed increasing amounts of the East Texas timber that had earlier been shipped to outside markets. A final change in the region's timber industry also reflected the fact that it was no longer simply a shipper of raw materials to other areas. This was the construction in 1937 of a large, modern, paper-manufacturing plant on the Houston Ship Channel at Pasadena. The plant of the Champion Paper and Fibre Company, an Ohio-based manufacturer of printing paper, became an important employer in the Pasadena area, and its construction symbolized two trends in the region's timber-related industry that continued in the post–World War II period: the production of finished products and the increasing role played by large, nationally oriented companies.[29]

The export of grain had not compared with that of either cotton or timber before World War I, but in the 1920s, Houston area promotors foresaw a great future for grain as a new regional export. They were right—at least for a while. After the construction of a large public grain elevator on the ship channel in 1926, shipments from Houston boomed. But the Great Depression brought an abrupt halt to this trade, as export markets collapsed. Indeed, Houston's grain elevator remained empty from 1932 to 1938, and other Gulf Coast elevators shared its fate.[30]

Rice shipments did not decline so drastically. The Sabine-Neches area continued to grow substantial crops of rice, and Beaumont in particular profited from its export. As the rice culture spread throughout the low-lying areas to the east and west of the Beaumont-Port Arthur area, the newly created port at Lake Charles, then a small Louisiana town some fifty miles east of Beaumont, experienced a boom in rice growing and shipping which made it a thriving new center of commerce.[31]

The emergence of Lake Charles as a port points out the importance to all regional industry of improvements in the coastal transportation system. The interwar years brought both a vast highway construction program and a significant upgrading of the shipping facilities throughout the region. Extensive dredging of the Neches River, completed in 1916, allowed Beaumont to expand its port greatly; Orange and Lake Charles also benefited from extensive work on their shipping lanes. The remainder of the region's major ports—Houston, Galveston, Port Arthur, and Texas City—all underwent steady and expensive improvements in this middle period. As such work greatly expanded their capacity to handle ocean-going cargoes, the gradual completion of an intracoastal canal connecting the ports of Texas with other Gulf Coast ports had a similar effect on shipping along the coast. Taken as a whole, these advances in transportation were an essential part of the economic expansion of the region during the heyday of refining.[32]

This was especially apparent in the case of the Port of Houston. The extraordinary expansion of shipping after the opening of the Houston Ship Channel in 1914 was testimony to its regional importance. As the channel attracted numerous new industries, shipping itself became a major area employer. As the total tonnage handled at the Port of Houston increased from 1.3 million in 1919 to almost 14 million in 1929 and finally to more than 27 million in 1941, it outdistanced its former rival, Galveston, and began to challenge New Orleans' historical leadership in Gulf Coast shipping.[33] By the 1930s, Houston was already the sixth largest port in terms of foreign commerce handled, behind only New York, New Orleans, Los Angeles, Baltimore, and Philadelphia. Because of its tremendous oil and cotton shipments, it ranked behind only New York and Los Angeles in total exports. Defenders of New Orleans' two-century-long reign as the center of Gulf Coast shipping disdainfully noted that much of the cargo handled by its brash, young challenger was merely "liquid cargo," but the vigilance displayed by the Port of New Orleans in attempting to combat its new rival belied this smugness.[34]

THE SHIFT IN POPULATION TO THE COASTAL CITIES

Houston's overall growth during this period was rapid enough to encourage such apprehension. Spurred by the industrial activity in the ship channel area, the city outstripped all other cities in Texas and throughout the South. By 1940, Houston ranked as the largest city in Texas, the second largest city in the South behind New Orleans, and the twenty-first largest in the entire nation. From 1910 to 1940, its population almost quintupled, a record of expansion exceeded among major cities in this period only by Los Angeles.

Other cities in the coastal region could not match this pace, and Houston grew in size relative to the Beaumont-Port Arthur area and to Galveston (see Table 3.11). In 1920, before the rapid industrial expansion along the Houston Ship Channel, the combined population of Beaumont and Port Arthur was about 45 percent of Houston's total, but by 1940 this percentage had plunged to 27. The shift in regional population to the east that had accompanied the initial development of oil before World War I had been reversed. At the same time, Galveston's decline relative to Houston continued as in the earlier years. Houston's ascendancy in the region could no longer be doubted, even by the most fervent boosters of the other coastal cities.

Led by the growth of its urban centers, the coastal region steadily pushed to the forefront in the larger section of eastern Texas and western Louisiana as the pull of new industrial jobs, especially in the 1920s, encouraged a steady migration from the interior to the coast. (See Table 3.12.) The cotton countries, which in 1910 had by far the largest population of any of the regions included in this table, grew very little in the next two decades and then registered an absolute loss in the 1930s. The timber and mixed regions fared a bit better, but neither grew as rapidly as the coastal region. The result was a dramatic relocation of sectional population during the heyday of refining. Thirty years of sustained, refinery-led development (1910–1940) helped double the coastal region's share of the total sectional population, increasing it from about 23 to about 46 percent and, in the process, encouraging a marked alteration in the distribution of population throughout an entire section of eastern Texas and western Louisiana.

Table 3.11. The Growth of Cities in the Refining Region: 1910-1940

	1910	1920	1930	1940	% Change 1910-1940
Houston	78,800	138,276	292,352	384,514	388
Beaumont	20,640	40,422	57,732	59,061	186
Port Arthur	7,663	22,251	50,902	46,104	502
Galveston	36,981	44,255	52,938	60,862	65
U.S.	91,972,266	105,710,620	122,775,046	131,669,275	43
U.S. Urban	41,998,932	54,157,973	68,954,823	74,423,702	77

Source: U.S. Census of Population, 1910-1940.

The primary economic magnet attracting people and resources to the upper Texas Gulf Coast during the middle years was the burgeoning petroleum industry. Its rapid growth tied the coastal region firmly into an expansive sector of the national economy while helping catapult Houston into the ranks of the nation's major cities. An impressive array of oil-related manufacturing constituted a growing industrial complex around which the regional economy developed, and the giant refineries on the Houston Ship Channel and in the Port Arthur area were at the

Table 3.12. Regional and Sectional Population Growth: 1910-1940

	1910	1920	1930	1940	% Change 1910-1940
Coast	251,700	389,850	651,050	871,134	246
Cotton	443,770	473,790	492,564	486,018	10
Mixed	152,988	174,569	184,172	202,424	32
Timber	261,800	303,921	308,484	330,953	26
Total	1,111,061	1,342,130	1,636,270	1,890,529	70
Total -Coast	859,361	952,280	985,220	1,019,395	19

See pages 20 and 21 above for an explanation of these divisions.

center of this complex. The construction and operation of these refineries as well as other aspects of the petroleum industry encouraged the growth of other related industries, the improvement of transportation, and the accumulation of a variety of commercial and financial capabilities. These, in turn, could potentially attract still more economic activity to the region.

This potential was not, however, fulfilled by the end of the middle period. The regional economy remained heavily reliant on its major export, petroleum products. Some observers interpreted this as proof that the region was essentially a "colonial" economy which exported raw materials to the industrialized Northeast while importing most of its finished goods from outside the region. These were indeed the characteristics of a colonial economy, but they were also the characteristics of a rapidly developing region which was only beginning to reach economic maturity. The impetus for growth generated by the petroleum industry had had only four decades—a very short time in developmental perspective—to work its way through the regional economy. Those in the region were not, after all, bewildered peasants unable to participate in the region's industrial expansion. Rather, decades of previous development and their American culture made them eager and able to claim a share of the benefits of development. If some eastern investors were reaping financial rewards via stocks and bonds, at least their investments were being used to build large, permanent manufacturing plants which bolstered the regional economy. Most of the on-going economic benefits generated by the expansion of these refineries and of other closely related branches of the oil industry accrued to the region, thus establishing a firm foundation upon which a more diversified and even more dynamic regional economy would be built in the post-World War II era. The growth pole was indeed in the process of reaching out and transforming the coastal economy. What perhaps resembled a "colony" when viewed as a stationary economy was, when seen in historical perspective, a dynamic economy moving very quickly from a heavy dependence on exports of crude oil toward industrial maturity.

NOTES

1. Sam Schurr and Bruce Netschert, *Energy in the American Economy, 1850-1975* (Baltimore, 1960), p. 39.

2. American Petroleum Institute, *Petroleum Facts and Figures*, 1971 Edition (Washington, 1971), p. 307. See, also, John B. Rae, *The American Automobile* (Chicago, 1965).

3. There are several useful histories of cracking. See especially, John Enos, *Petroleum Progress and Profits: A History of Process Innovation* (Cambridge, Mass., 1962). Cracking is the

general name used to describe a variety of refining processes which transform fuel oil into gasoline by splitting large hydrocarbon molecules into smaller ones. High temperature, pressure, and catalysts have been used in cracking.

4. *Oil and Gas Journal,* August 1934, Supplement, p. 199. See, also, Walter Miller and Harold Osborn, "History and Development of Some Important Phases of Petroleum Refining in the United States," in Albert Duncan (ed.), *The Science of Petroleum,* vol. 2 (New York, 1938), pp. 1466-1478.

5. For a discussion of the major innovations, see John Enos, *Petroleum Progress and Profits,* pp. 131-224. For statistics on cracking capacity, see American Petroleum Institute, *Petroleum Facts and Figures,* 1971 Edition, p.163.

6. American Petroleum Institute, *Petroleum Facts and Figures,* 1971 Edition, p.154.

7. The capacity for Standard is taken from Ralph and Muriel Hidy, *Pioneering in Big Business, 1882-1911* (New York, 1955), pp. 410-417; those for Jefferson County from the United States Bureau of Mines (hereafter USBM), Information Circular 6485 (May, 1931). This is one of a yearly series since 1918 of *Petroleum Refineries in the United States,* which lists information for each individual refinery as of January first for each year.

8. Carl Coke Rister, *Oil! Titan of the Southwest* (Norman, Oklahoma, 1949) discusses the discovery of oil throughout the Southwest after Spindletop.

9. U.S. Department of Interior, Bureau of Mines, "Reports of Investigations," numbers 2964 and 3038.

10. The overland distances from the major Southwestern fields to the Gulf Coast range from 0 to 750 miles; these fields are 1000 to 2000 miles from the East Coast.

11. Henrietta Larson and Kenneth Porter, *History of Humble Oil and Refining Company* (New York, 1959), p. 216.

12. Hearings Before the Temporary National Economic Committee, Part 14, Petroleum Industry, Section 2, pp. 8630-8631.

13. As of January 1, 1940, the capacities of the ten largest refineries in the U.S. were as follows (in barrels per day):

*1. Humble Oil—Baytown—140,000
*2. Texaco—Port Arthur—135,000
*3. Gulf—Port Arthur—125,000
4. Jersey Standard—Bayonne, New Jersey—124,000
5. Standard of Indiana—Whiting, Indiana—119,000
*6. Mobil—Beaumont—100,000
7. Jersey Standard—Baton Rouge—100,000
8. Standard of California—El Segundo, Cal.—100,000
Richmond, Cal.—100,000
*10. Standard of Indiana—Texas City—94,000
*Located on the upper Texas Gulf Coast.

Source: U.S. Department of Commerce, "U.S. Petroleum Refining: War and Postwar" (Washington, 1947), appendix 1.

14. *Oil and Gas Journal,* December 10, 1914, p. 22; September 7, 1916, p. 2. For an account of the early history of the Mexican oil industry, see Pan American Petroleum & Transport Company, *Mexican Petroleum* (New York, 1922). For a recent account of the disputes between the U.S. and British oil companies and the Mexican government, see Lorenzo Meyer, *Mexico and the United States in the Oil Controversy, 1917-1942* (translated by Muriel Vasconcellos) (Austin, 1977).

15. Henrietta Larson and Kenneth Porter, *History of Humble Oil,* pp. 193-218.

16. U.S. Bureau of Mines, Information Circular 7161, pp. 10–20.

17. Submittal to the T.N.E.C. by Texaco, "Port Arthur Works," in Texas Company Archives, New York.

18. U.S. Department of Commerce, "U.S. Petroleum Refining: War and Postwar," appendix 1.

19. Beaumont Chamber of Commerce. "Beaumont, Texas: Statistical Review of the Progress for 1925" (Beaumont, 1925), pp. 28–31. This booklet is located in the Barker Library at the University of Texas, Austin.

20. This quote from the December 1939 issue of *Fortune* is repeated in W.P.A.—American Guide Series, *Houston: A History and Guide* (Houston, 1942).

21. U.S. Bureau of the Census, *Census of Manufactures, 1930* (Washington, 1930), p. 516.

22. For a summary of this production, see *Houston in Brief,* 1932 edition (E. R. Bogy, 1932), pp. 67–69. One copy is located in the Library of Congress, Washington, D.C. See, also, Carl Coke Rister, *Oil! Titan of the Southwest. Beaumont Enterprise,* October 5, 1941, p. 2F lists 185 producing oil fields within 150 miles of Houston as of 1941.

23. U.S. Bureau of the Census, *Census of Manufactures, 1939* (Washington, 1940), p. 997.

24. For reports on regional refinery construction, see *Beaumont Enterprise,* September 14, 1960, p. 3; January 6, 1938, p. 5; October 18, 1936, p.1.

25. W.P.A., *Houston: A History and Guide,* p. 159. *Beaumont Enterprise,* October 5, 1941, p. 2F.

26. David McComb, *Houston, The Bayou City* (Austin, 1969), p. 112.

27. *Houston in Brief,* 1932 edition (Houston, 1932), pp. 40–41. See, also, Houston Cotton Exchange and Board of Trade, *Fifty Years a Cotton Market* (Houston, 1924).

28. U.S. Department of Interior, Bureau of Mines, *Pollution by Oil of the Coast Waters of the United States,* appendix D. (Washington, 1923), pp. 434–435. See, also, Hamilton Easton, "The History of the Texas Lumbering Industry," unpublished doctoral dissertation, University of Texas—Austin, 1947.

29. *Beaumont Enterprise,* October 10, 1954, Section VII, p. 2.

30. Marilyn Sibley, *The Port of Houston* (Austin, 1968), p. 175.

31. *Beaumont Enterprise,* August 3, 1926, p. 30A. See, also, Joseph Clark, *The Texas Gulf Coast: Its History and Development,* vol. 2 (New York, 1955), p. 73.

32. For a good account of many of these improvements, see Lynn Alperin, *Custodians of the Coast: History of the United States Army Engineers at Galveston* (Galveston, 1977).

33. For export figures, see U.S. Army Corps of Engineers, *The Port of Houston* (Washington, 1936); *The Port of Galveston* (Washington, 1936); and *The Port of New Orleans* (Washington, 1932).

34. Marilyn Sibley, *The Port of Houston,* pp. 161–162.

Chapter IV

The Modern Era (1941–1960s): Refining in a Maturing Regional Economy

During the two decades prior to World War II, the phrase "refining region" more aptly characterized the economy of the upper Texas Gulf Coast than at any other period in its history. In the modern era which commenced with the onset of the war, however, this phrase became less descriptive as other industries challenged the economic dominance of refining and as the region became increasingly less distinct from the remainder of the rapidly developing Gulf Coast. In the post-war decades, the giant refineries continued to expand, but other industries—most notably the production of petrochemicals—also took root and flourished in an economic environment made fertile by the previous forty years of oil-based development. Although the resulting industrial growth spread throughout the area from Corpus Christi to New Orleans, Houston nonetheless remained the focal point of coastal expansion. Its ascent to prominence among the nation's cities in the 1960s and 1970s well symbolized the coming of economic age of both the city and the surrounding region.

REFINING IN THE AUTOMOBILE AGE

World War II brough an urgent need for increased supplies of numerous petroleum products. High octane aviation fuel, gasoline and diesel fuel for mechanized warfare, petroleum-based synthetic rubber, and other essential goods flowed from the nation's refineries to the battlefields. Rapid growth and technological change accompanied the refining industry's successful response to the difficult demands placed on it by the war effort, and the industry emerged from the war with 15 percent more refining capacity and 45 percent more cracking capacity than it had possessed in 1941.[1]

This additional capacity helped the industry readjust quickly to a period of rising consumer demand after the war. The most important source of this surge in demand was the increasing use of gasoline in automobiles: from 1946 to 1969 the total number of motor vehicles registered in the United States rose from approximately 34 million to almost 100 million. In the same period, total miles driven in the U.S. more than tripled while the average number of miles traveled per gallon of fuel actually declined by 9 percent.[2] These figures reflected the growth of the suburbs, which meant a greater reliance on the automobile for long distance commuting, the construction of an extensive interstate highway system, which encouraged the use of motor vehicles for the movement of passengers and freight between cities, and the growth in the average weight and power of the passenger car, which increased fuel consumption.[3] More cars were thus being driven more miles while requiring more fuel per mile, and to meet the rising demand for gasoline, the refining industry expanded steadily throughout the 1940s, 1950s, and 1960s.[4]

The demand for other refined petroleum products also rose, often dramatically. Two relatively new products—fuel for jet airplanes and feedstock for petrochemical production—found expanding markets in the modern era. The use of asphalt and other petroleum-based road building material showed substantial growth amid the highway construction boom of the 1950s and 1960s. Finally, the demand for traditional products such as fuel oil and kerosene continued to increase. As the use of gasoline surged, as new uses for petroleum emerged, and as oil continued to replace coal and other sources of energy for such functions as heating homes and providing fuel for industrial processes, the total domestic demand for petroleum products soared from about 5.3 million barrels per day in 1946 to more than 13.3 million in 1969 (see Table 4.1).

Even this rapid rate of increase was far surpassed by the rate of growth in the use of petroleum throughout the remainder of the world. Western European countries showed the same passion for the automobile in the 1950s and 1960s that the United States had begun to exhibit in the 1920s; almost 63 million motor vehicles were registered in Western Europe in 1969, a figure that was fifteen times the total registered in 1946.[5] Similar, though less dramatic, increases in the use of gasoline and

Table 4.1. Total U.S. Demand for Petroleum Products, 1941–1968 (in thousands of barrels per day)

Year	Total Demand	Percent Increase
1941	4,369	-
1942	4,293	-1.8
1943	4,579	6.2
1944	5,134	10.8
1945	5,358	4.2
1946	5,331	-0.5
1947	5,902	9.7
1948	6,143	3.9
1949	6,130	-0.2
1950	6,812	10.0
1951	7,463	8.7
1952	7,712	3.2
1953	8,005	3.7
1954	8,115	1.4
1955	8,827	8.1
1956	9,209	4.1
1957	9,386	1.9
1958	9,358	-0.3
1959	9,662	2.9
1960	9,863	2.0
1961	9,980	1.2
1962	10,403	4.1
1963	10,759	3.3
1964	11,017	2.3
1965	11,489	4.1
1966	12,048	4.6
1967	12,584	4.3
1968	13,350	5.7

Source: American Petroleum Institute, *Petroleum Facts and Figures*, 1971 edition (Washington, 1971), pp. 283–287.

other petroleum products occurred throughout the world in the post-war era. The total demand for refined petroleum products in noncommunist countries reflected the tremendous expansion of the industry: in 1950 this demand was approximately 8.5 million barrels per day; in 1969 it had risen to more than 36.6 million barrels daily.[6] American oil companies took a leading role in supplying this worldwide demand, and in the process they became increasingly international in outlook.

The sustained expansion of refining output required to meet the increased domestic and international demand in the post-World War II era is difficult to envision. A benchmark introduced in the last chapter—the old Standard Oil combination which operated a refining capacity of slightly more than 300,000 barrels per day before its dissolution in 1911—again provides an instructive measure of the growth of the refining industry in the modern era. By 1946, the refining capacity of "the Octopus" was already exceeded by six indiviudal companies in the United States. By 1971, only sixty years after the break-up of Standard Oil, six separate refineries on the Gulf Coast processed more oil per day than had the entire Standard Oil combination during its peak production. Indeed, in the single year of 1971, refineries on the Texas-Louisiana Gulf Coast refined more oil than had been refined by Standard from its creation in 1882 until its dissolution in 1911. More than 120 "octopi" would have been required in 1969 to provide the refined petroleum consumed in the non-communist world.[7]

In the modern era, as in earlier years, the Gulf Coast shared disproportionately in the growth of refining required to meet the steady and at times spectacular expansion in the demand for petroleum products. Table 4.2 shows the continued predominance of the Gulf Coast among all national refining areas, and it also reveals a gradual expansion in the percentage of total refining capacity located on the Gulf Coast.

As the largest refining region in the nation, the Gulf Coast remained an important supplier of refined petroleum products to the urban centers of the northeastern United States. Table 4.3 records the increase in tanker shipments from the Gulf to the East Coast, but these statistics also reveal a decline in the percentage of Gulf Coast production being shipped by water to the East. This relative decline reflected the growing use of pipelines for long-distance shipping of petroleum products in the modern ear. In addition, the rapidly industrializing southwestern United States absorbed a greater proportion of the region's refined goods, thus diverting some of the traditional shipments from the eastern markets. Finally, as first Venezuelan and then Middle Eastern oil fields came to replace the fields of the southwestern United States as the most important sources of crude, the Gulf Coast began to lose one of its comparative advantages over other refining regions.[8] Such changes weakened, but did not break, the region's traditionally strong ties to

Table 4.2. Changing Geographical Location of Operating Refining Capacity in the U.S. 1941–1976 (Percentage of National Capacity in Each Region)

Year	1	2	3	4	5	6	7	8	9
1941	27.9	15.4	3.6	15.9	9.0	5.3	1.9	2.6	18.3
1947	29.4	15.2	3.4	16.5	8.0	5.1	1.4	3.1	18.1
1951	30.5	15.2	2.8	18.0	8.5	3.7	1.2	3.8	16.2
1956	32.2	12.2	2.4	18.9	8.5	3.5	1.2	3.8	15.8
1961	30.5	14.8	2.4	19.3	8.4	3.8	1.4	3.7	15.7
1966	32.5	12.5	2.4	19.3	8.2	3.6	1.5	4.1	15.8
1971	36.2	10.6	1.7	19.8	7.3	3.3	1.4	3.7	15.9
1976	35.9	10.0	1.8	19.8	7.3	3.6	1.5	4.4	15.5

1. Gulf coast of Texas and Louisiana.
2. East coast - contains all Atlantic coast states and the eastern portions of New York and Pennsylvania, including New York City and Philadelphia.
3. Appalachian - Western Virginia, the remainder of New York and Pennsylvania and the eastern portion of Ohio, including the Cleveland area.
4. Indiana, Illinois, and Kentucky - plus the remainder of Ohio and Tennessee, Michigan and Minnesota, Wisconsin, North and South Dakota after 1951.
5. Oklahoma, Kansas, Missouri - also Nebraska and Iowa.
6. Texas inland - all of Texas minus the Gulf coastal region.
7. Arkansas, Louisiana inland - does not include Louisiana Gulf coast.
8. Rocky Mountain - Montana, Idaho, Wyoming, Utah, Colorado, and New Mexico.
9. West coast - primarily California.

For source and further explanation of districts used, see Table 3.4, page 66.

markets on the Atlantic seaboard. Although these markets continued to consume large amounts of the products of the Gulf Coast refineries, they were no longer as important to the region as they had been before World War II. In the modern era, the Gulf Coast refineries served more diverse markets in a petroleum industry that was becoming increasingly international.

Table 4.3. Water Shipments of Petroleum Products from Gulf Coast Ports to Atlantic Ports 1941–1966 (Thousands of 42-gallon barrels) Total Shipments by Products and Shipments as a Percentage of Gulf Coast Production

Year	Gasoline	Percent of Total Gulf Coast Production	Fuel Oils[1]	Percent of Total Gulf Coast Production	Crude	All Petroleum Products
1941	130,534	69	118,543	71	147,288	430,769
1946	123,967	55	124,037	57	189,961	481,690
1951	180,126	52	190,332	67	194,913	629,687
1956	227,938	49	247,434	68	168,793	710,767
1961	241,860	50	236,136	78	175,481	719,478
1966	182,475	31	179,865	53	156,001	622,250

[1]Includes Gas Oil and Distillate Fuel Oil and Residual Oil.

Source: *Petroleum Facts and Figures*, 1971, p. 262.

THE SPREAD OF REGIONAL REFINING IN THE MODERN ERA

Some of the most important changes affecting the refining region in the post-World War II period began during the war. The Gulf Coast, which was less vulnerable to enemy attack than other refining centers, was a favored site for the expansion of refining required by the war effort. The region's natural locational advantages were enhanced by the government's decision to construct the first pipelines from the Gulf Coast to the East Coast.[9] The government also met the exigencies of war by investing more than half a billion dollars at or adjacent to existing refineries in an effort to assure adequate supplies of certain essential war materials. Sixty percent of this total was spent on the Gulf Coast, including more than $100 million for facilities to produce high octane aviation fuel and almost $150 million to build facilities for the manufacture of synthetic rubber, which was urgently needed to replace the supplies of natural rubber made inaccessible by the war. Such government investments were more than matched by private funds, which accounted for more than three-fourths of the one billion dollars invested nationally in high octane gasoline facilities. This combination of public and private investment boosted the region's overall refining capacity, upgraded much of the equipment used in the refineries, and established a major new refining-related industry, the production of petroleum-based synthetic rubber.[10]

The war also brought a significant geographical shift in the location of refineries along the Gulf Coast. One of the major refinery construction projects completed with government funds during the war was the large, technologically advanced Cities Service Oil Company plant at Lake Charles, Louisiana, some forty miles east of Port Arthur. The Continental Oil Company also build a small refinery near the same city. The subsequent growth of these two plants and the addition of other petroleum-related manufacturing in the post-war period marked the emergence of Lake Charles as an important new center of oil refining and petrochemical production.[11]

The older refining centers on the Gulf Coast also expanded during and after the war. The overview of the growth of refining in the modern era presented in Table 4.4 reveals several significant trends. During the thirty-five years between 1941 and 1976, the refining capacity located on the Gulf Coast expanded by more than 360 percent. Much of this increase occurred in the upper Texas Gulf Coast, where daily capacity grew from one million barrels to more than three million. Within this region, the Port Arthur area kept pace with the Houston area until the mid-1960s, after which it began to fall behind in total capacity. In the early 1970s, Jefferson County's share of the total refining capacity of the United States dipped below 10 percent for the first time since early in the century. The continued expansion of the capacity of the Houston area allowed it to outstrip Port Arthur, the historical center of Gulf Coast refining.

Despite the impressive growth in refining capacity on the upper Texas Gulf Coast, its percentage of total Gulf Coast capacity declined sharply from almost 80 percent in 1941 to less than 60 percent in 1976. This shift reflected the fact that numerous new refineries were being built in areas that were not crowded from previous construction. The 1940s boom in refinery construction near Lake Charles was matched by a similar boom near Corpus Christi in the 1950s. An even greater geographical shift followed in the 1960s and 1970s, when the New Orleans area emerged as a major refining center. Shell and Tenneco led the way by expanding their existing plants, which originally had been built near

Table 4.4. Operating Crude Capacity on the Texas-Louisiana Gulf Coast 1941-1976 (42-gallon barrels per day)

Year	Total	Percent Increase	Beaumont-Port Arthur	Percent of Gulf	Houston	Percent of Gulf
1941	1,167,400	50.8	450,600	38.6	472,800	40.5
1947	1,566,900	34.2	580,800	37.1	539,300	34.4
1951	2.042,350	30.3	718,200	35.2	740,200	36.2
1956	2,700,700	32.2	845,900	31.3	925,600	34.3
1961	2,933,540	8.6	958,540	32.7	946,800	32.3
1966	3,211,600	9.5	1,036,000	32.3	1,085,350	33.8
1971	4,303,000	34.0	1,232,700	28.6	1,453,500	33.8
1976	5,400,000	25.5	1,231,600	22.8	1,941,000	35.9

Year	Corpus Christi	Percent of Gulf	Louisiana Gulf	Percent of Gulf
1941	106,500	9.1	137,500	11.8
1947	94,500	6.0	352,300	22.5
1951	93,750	4.6	490,200	24.0
1956	225,000	8.3	704,200	26.1
1961	251,000	8.6	777,200	26.5
1966	245,000	7.6	845,250	26.3
1971	341,000	7.9	1,275,800	29.6
1976	471,200	8.7	1,756,200	32.5

Source: U.S. Bureau of Mines, Petroleum Refining Information Circulars, 1941-1976.

New Orleans in the 1920s during the brief period of expansion brought about by the availability of Mexican oil. In the 1960s and 1970s, Gulf Oil and Texaco constructed large new plants along the Mississippi River, helping boost the New Orleans area's capacity from 230,000 barrels per day in 1966 to more than 930,000 barrels per day in 1976, a figure that represented more than 17 percent of the total capacity on the Gulf Coast.[12]

As large refineries thus spread along the coast from Corpus Christi to New Orleans, the upper Texas Gulf Coast became less distinct as a "refining region" separate from this larger area. Before the modern era, the region from Port Arthur to the Houston Ship Channel had contained all of the major Gulf Coast refineries except the Exxon plant at Baton Rouge. By the 1970s, however, the region was simply the largest and oldest of numerous refining centers on the Texas-Louisiana Gulf Coast.

In part responsible for this geographical shift was a change in the pattern of refinery ownership. In the 1920s and 1930s, the seven largest oil companies active on the Gulf Coast controlled from 80 to 90 percent of its refining capacity. In the late 1930s, however, this percentage began to decline. This trend continued after World War II, and by 1961 the seven largest companies controlled about sixty-seven percent of Gulf Coast capacity. (See Table 4.5.) Most of the remaining capacity was owned by a group of refiners that were somewhat smaller than Gulf Oil, Texaco, Exxon, Mobil, Shell, Amoco, and Sinclair—the companies that had earlier controlled most regional refining capacity. Corpus Christi was a favored location for the refineries of growing companies such as Pontiac Refining Company and Suntide Refining Company. Several other relative newcomers to the Gulf Coast, notably Eastern States Petroleum Company, Crown Central Petroleum, and Republic Oil and Refining Company, operated plants along the Houston Ship Channel. The expanding regional role of such companies in the modern era was the single most significant alteration in the pattern of ownership of Gulf Coast refining since the initial emergence of Gulf Oil and Texaco.

These new companies were small only in comparison to the largest oil companies. They often grew into vertically integrated oil companies serving national and even international markets, and their regional refineries generally represented substantial investments. In short, these were not "mom and pop" operations at the mercy of the large oil companies, but rather dynamic young companies that were often potentially capable of breaking into the ranks of the largest companies.

Table 4.5 reflects in part the dictates of changing techniques of production. The growing use of modern cracking techniques encouraged the steady expansion of the average size of the giant coastal refineries, thereby attracting to the region the vast new investments required to complete this expansion. The resulting economic benefits were, however, somewhat offset by another technological change in the refineries, the rapid introduction of automation. In the modern era, increases in refining capacity no longer translated into increases in total employment, and this change gradually weakened the strongest direct linkage between the refineries and the regional economy. Although the trend

Table 4.5. Concentration of Refinery Ownership on the Gulf Coast

NUMBER OF COMPANIES IN OPERATION:

	1918	1921	1926	1931	1936	1941	1947	1951	1946	1961
Largest	4	5	5	6	7	7	7	7	7	7
Large	3	5	5	3	3	8	10	12	14	14
Small	6	9	6	5	10	15	14	5	6	8

PERCENT CAPACITY OWNED:

	1918	1921	1926	1931	1936	1941	1947	1951	1956	1961
Largest	85	77	79	90	88	79	75	75	69	67
Large	11	17	15	8	6	14	20	24	30	31
Small	5	6	6	2	6	8	5	1	1	2
Largest-Large	96	94	94	98	94	93	95	99	99	97

AVERAGE CAPACITY PER COMPANY: (thousands of barrels per day)

	1918	1921	1926	1931	1936	1941	1947	1951	1956	1961
Largest	45	40	71	101	98	132	170	216	265	279
Large	8	9	14	18	16	20	31	41	58	65
Small	2	2	5	3	4	6	6	3	5	6
Major Refinery of Largest	42	31	53	76	74	102	135	172	209	223

Source: Yearly Petroleum Refining Information Circulars of the U.S. Bureau of Mines.

Largest - Seven largest refiners active on the Gulf Coast.
Large - The next twenty refiners by size of refining capacity.
Small - All others.

For a detailed list of the companies grouped in each category, see Appendix 2.

toward fewer man hours per barrel of product had characterized refining throughout most of its history, its impact became most noticable in the 1950s and 1960s. In the coastal region, the Port Arthur-Beaumont area was most severly affected, for although its refineries continued to grow in the modern era, their work forces declined sharply. Unlike the diversifying economy of the Houston area, Port Arthur and Beaumont did not develop other major industries to take up the slack caused by the declining employment in refining, and the growth of their population and industrial output slowed dramatically in comparison to earlier years.[13]

REFINING IN A DIVERSIFYING ECONOMY

Increasing diversification was a defining characteristic of the regional economy after 1941, but petroleum refining nonetheless retained a prominent economic role in the region. In supplying refined petroleum products to national and international markets, the large coastal refineries continued to employ a substantial portion of the region's industrial labor force while accounting for much of its total value added by manufacturing (see Tables 4.6, 4.7, and 4.8). Due to the growth of other industries, however, petroleum refining no longer dominated the regional economy as in earlier decades.

Many of the new industries were closely tied to refining. By far the most important of these, the production of petrochemicals, grew rapidly throughout the modern era, giving the regional economy a tremendous boost. Several other economic pursuits that were directly and indirectly encouraged by the continuing expansion of the oil industry also contributed to the prosperity of the region. The composition of this oil-linked industrial complex is suggested by Table 4.6, which also shows its changing magnitude. The ties among these industries will be analyzed in Chapter V, but the remainder of this section will present an overview of their growth. The expansion of such oil-related industries, the evolution of traditional regional industries such as cotton, timber, and grain production, and the emergence of several new industries, most notably the establishment of the space program's headquarters near Houston, all added a measure of diversity to a regional economy still greatly influenced by the petroleum industry.

In light of the increasing importance of petrochemicals in the regional economy, a brief description of the history of this industry is in order. "Petrochemicals" is a generic term used to include a variety of intermediate and finished products derived from the hydrocarbons which make up petroleum and natural gas.[14] As recently as 1920, only two petrochemical companies were active in the United States. A chemical company, Union Carbide, and a petroleum company, Standard of New Jersey, pioneered in the development of new, petroleum-based products in the 1920s and 1930s, but the petrochemical industry expanded slowly

Table 4.6. Manufacturing in the Houston Area, 1947–1972

	Total of all Manufacturing	Petroleum and Coal Products[a] (29)	Chemicals and Allied Products (28)	Primary Metals Industries (33)	Fabricated Metals Industries (34)	Machinery Except Electrical[b] (35)
			Manufacturing Employment			
1947	58,606	10,955 (19%)	4,500 (8%)	3,949 (7%)	4,877 (8%)	11,954 (20%)
1954	77,782	12,028 (15%)	9,071 (12%)	5,293 (7%)	7,799 (10%)	15,525 (20%)
1958	89,294	12,807 (14%)	9,798 (11%)	6,132 (7%)	11,454 (13%)	15,683 (18%)
1963	108,585	10,549 (9%)	15,753 (15%)	9,723 (9%)	13,205 (12%)	15,351 (14%)
1967	138,100	10,600 (8%)	17,500 (13%)	13,500 (10%)	18,800 (14%)	19,400 (14%)
1972	159,700	9,100 (6%)	22,400 (14%)	10,100 (6%)	22,600 (14%)	26,800 (17%)

Value Added by Manufacturing (millions of dollars)

1947	385	90 (23%)	49 (13%)	17 (4%)	28 (7%)	72 (19%)
1954	869	153 (18%)	198 (23%)	52 (6%)	57 (7%)	183 (21%)
1958	1,154	196 (17%)	262 (23%)	76 (7%)	101 (9%)	179 (16%)
1963	1,918	340 (18%)	598 (31%)	133 (7%)	133 (7%)	210 (11%)
1967	2,873	549 (19%)	836 (29%)	234 (8%)	254 (9%)	315 (11%)
1972	4,180	458 (11%)	1,362 (33%)	207 (5%)	410 (10%)	543 (13%)

Source: U.S. Census of Manufactures, Area Statistics: 1947, p. 589; 1954, pp. 142-16, 142-17; 1958, p. 42-17; 1963, p. 44-19; 1967, pp. 44-20, 44-21; 1972, pp. 44-21, 44-22, 44-23.

Note: The numbers in parentheses are the two-digit groupings of the Standard Industrial Classification used in the Census of Manufactures. I have used the Census of Manufactures designation for the Houston Standard Metropolitan Area to define the "Houston Area." In 1947, 1954, and 1958, this included Harris County; in 1963, Harris, Brazoria, Fort Bend, Liberty, and Montgomery counties; in 1967, Harris, Fort Bend, Liberty, and Montgomery; in 1972, Harris, Fort Bend, Liberty, Montgomery, Brazoria, and Waller counties.

[a] For the Houston area in the years listed, the four-digit grouping 2911, "petroleum refining," made up more than 95 percent of this two-digit category.

[b] For the Houston area in the years listed, the four-digit grouping 3532, "oil field machinery and tools," made up from 60 to 85 percent of this two-digit category.

Table 4.7. Manufacturing in the Port Arthur-Beaumont Area,[a] 1947–1972

	Total of all Manufacturing	Petroleum and Coal Products (29)	Chemicals and Allied Products (28)	Primary Metals Industries (34)	Fabricated Metals Industries (34)	Machinery Except Electrical (35)
			Manufacturing Employees			
1947	23,181	16,743 (72%)	950 (4%)	269 (1%)	1,235 (5%)	276 (1%)
1954	25,445	17,653 (69%)	1,558 (6%)	259 (1%)	1,058 (4%)	917 (4%)
1958	26,264	17,829 (68%)	2,251 (9%)	na	1,628 (6%)	na
1963	24,893	14,659 (59%)	3,127 (13%)	na	1,215 (5%)	714 (3%)
1967	33,200	13,100 (39%)	8,200 (25%)	400 (1%)	2,900 (9%)	500 (2%)
1972	37,600	14,100 (38%)	8,900 (24%)	na	2,200 (6%)	800 (2%)

Value Added by Manufacturing (millions of dollars)						
1947	178	134 (75%)	16 (9%)	1 (1%)	6 (3%)	na
1954	223	151 (67%)	34 (15%)	2 (1%)	8 (4%)	5 (2%)
1958	259	165 (64%)	40 (15%)	na	15 (6%)	na
1963	539	384 (71%)	89 (17%)	na	15 (3%)	7 (1%)
1967	1,166	772 (66%)	269 (23%)	6 (1%)	32 (3%)	9 (1%)
1972	1,034	367 (35%)	439 (42%)	na	39 (4%)	17 (2%)

Source: U.S. Census of Manufactures, Area Statistics: 1947, p. 591; 1954, p. 142-18; 1958, p. 42-19; 1963, pp. 44-21, 44-22; 1967, p. 44-17; 1972, p. 44-18.

[a] I have used the Census of Manufactures designation for the Beaumont-Port Arthur Standard Metropolitan Area to define the "Port Arthur-Beaumont Area." In 1947, 1954, 1958, 1963, this included only Jefferson County; in 1967, Jefferson and Orange counties; in 1972, Jefferson, Orange, and Hardin counties.

Table 4.8. Percentage of Total Value Added by Manufacturing Contributed by Selected Industrial Activities in the Houston and Port Arthur-Beaumont Areas

	Total of all Manufacturing (in millions of dollars)	Petroleum and Coal Products (29)	Chemicals and Allied Products (28)	Primary Metals Industries (33)	Fabricated Metals Industries (34)	Machinery Except Electrical (35)
1947	563	40%	12%	3%	6%	13%
1954	1,092	28%	21%	5%	17%	6%
1958	1,413	26%	21%	5%	8%	13%
1963	2,457	29%	28%	5%	6%	9%
1967	4,039	33%	27%	6%	7%	8%
1972	5,214	16%	35%	4%	9%	11%

Source: See Tables 4.6 and 4.7.

until World War II, which radically altered its rate of growth. The war brought a sudden and vital need for the production of synthetic rubber and other products derived from petroleum, and a massive construction program financed by both public and private funds followed. After the war, the petrochemical industry continued to forge forward, and it remained one of the fastest growing major industries in the nation until the early 1960s. Throughout this period it grew by ten to twenty percent annually. In addition to this extraordinary expansion, an *Oil and Gas Journal's* special report on refining and petrochemicals in 1959 cited another important characteristic of its history in the modern era: "Perhaps the outstanding single development (in the petrochemical industry) was a geographical one—the postwar build-up of petrochemical facilities along the Gulf Coast."[15]

This assertion finds substantiation in an extensive study of the petrochemical industry in Texas completed in 1973. Although eight small petrochemical companies were active on the Texas Gulf Coast before 1940, the big departure in regional production came with World War II. In the 1940s, fifteen new petrochemical companies entered the region; in the 1950s, twenty-one; in the 1960s, seventeen. The results of this expansion were evident in a survey of the petrochemical industry on the Texas Gulf Coast in 1970 which revealed that this area produced about 40 percent of all U.S. petrochemicals and 80 percent of the nation's supply of synthetic rubber.[16]

Petrochemicals were the catalyst for a second spurt of regional growth in the modern era, which repeated in many ways the refining-led expansion of the period from 1901 to the 1930s. By 1970, six billion dollars had been invested in the petrochemical industry in Texas. In that year its value added reached 2.3 billion dollars, it employed 47,700 workers, and the value of all its shipped goods exceeded 4.3 billion dollars.[17] Most of this expansion occurred in the coastal area from Corpus Christi to Port Arthur, and—as shown in Tables 4.6 and 4.7 above—the economy of the upper Texas Gulf Coast in general and the Houston area in particular felt the impact of the surge in petrochemical production.

These tables show that "chemicals and allied products" even displaced "petroleum and coal products" as the largest manufacturing activity in the region, at least as measured by value added.[18] The significance of petrochemical production for the regional economy was not, however, that it displaced petroleum refining as the dominant regional industry, but rather that it increased the importance of the refineries by raising the value of their products. Most petrochemical production utilized production processes that were in a sense extensions of the cracking techniques earlier used to increase the value of petroleum products by recovering more gasoline from each barrel of crude oil refined. Petrochemistry further enhanced the value of the hydrocarbons in oil and natural gas by transforming them into a variety of even more valuable products. The maze of pipelines carrying products back and forth between the regions' oil refineries and its petrochemical plants was only the most obvious of many direct ties between the traditional bellwether industry and its modern "rival."

Because the large Gulf Coast refineries were the primary source of the "feedstock" used as the basic raw material for petrochemical production, the spread of petrochemical plants along the Gulf Coast followed the geographical pattern previously established by petroleum refining.[19] Indeed, in the modern era all of the major refiners on the upper Texas Gulf Coast constructed their own facilities for processing petrochemicals, usually adjacent to or inside of their existing refineries. Numerous other petrochemical plants grew outside of the direct control of the large refiners but remained dependent on them for much of their feedstock, and this dependence encouraged the selection of sites convenient to existing refining centers.

Jefferson County was thus one logical location for large-scale petrochemical production. Large facilities were built adjacent to the refineries of Gulf Oil and Texaco in Port Arthur and Mobil in Beaumont, and several other large new plants were constructed in the "mid-country" area between the two cities. During World War II, the small town of Port Neches was chosen as the site of the Neches Butane Products Company, which became the largest producer of butadiene, an essential ingredient in the production of synthetic rubber, in the world. Two large snythetic rubber plants were then constructed adjacent to Neches Butane, and several substantial petrochemical plants also grew nearby in the 1950s. During the same period, large plants were constructed near Beaumont by DuPont and by Goodyear. The area around Port Arthur witnessed similar expansion, and much of it spilled over the Jefferson County boundary into neighboring Orange County, where an impressive "chemical row" grew to include major manufacturing facilities of DuPont, Spencer Chemical Company, and Allied Chemical. By the 1960s and 1970s, petrochemical production had become roughly equal to petroleum refining as a manufacturing activity in the Port Arthur-Beaumont area (see Table 4.7), and the two industries together dominated that area's economy in the same way that refining had done in its heyday.[20]

Petrochemicals became even more important in the Houston area, for there they grew to surpass petroleum refining (see Table 4.6). The Houston Ship Channel area became a haven for petrochemical production. Numerous nationally active chemical and synthetic rubber companies and all of the major refiners previously active in the area constructed plants there. Down the ship channel on Galveston Bay, several more large plants were built near the petroleum refineries at Texas City. Farther south of Houston, in a coastal area that had not previously housed refineries, a massive petrochemical complex grew up around Brazosport. Dow Chemical Company signalled this significant geographical shift in the location of regional manufacturing in 1940 when it built a plant designed to recover magnesium from sea water. Dow subsequently expanded its investment near Brazosport, creating a vast industrial complex which spurred development along the coast south of Houston. Indeed, throughout the Houston area the rapid expansion of petrochemical production was perhaps the primary industrial stimulus for sustained economic growth in the modern era. The construction and the operation of these petrochemical plants thus greatly amplified the economic impact of the area's refineries.

The production of petrochemicals was far and away the largest new regional industry to emerge in the modern era, but it was by no means the only source of diversification. Another important new industry, the production of primary metals, marked the growing economic maturity of the region. Armco Steel led the way with the construction of a large steel mill at Galena Park, east of Houston, in 1941. The subsequent expansion of this plant and the completion of a large-diameter pipe mill adjacent to it made this steel producing complex a major employer in the Houston area. As will be discussed in more detail in the following chapter, other primary metals producers joined Armco on the Gulf Coast. These included Dow Chemical's magnesium recovery plant near Brazosport and Alcoa's modern aluminum producing facility at Port Comfort, down the coast about one hundred miles south of Houston. Other metals companies went still a hundred miles farther south, building plants in the Corpus Christi area. Taken as a whole, these primary metal producing plants gave an added boost to the post-World War II economy of the Texas Gulf Coast. By 1955, only fourteen years after Armco's initial construction near Houston, more than $250 million had been invested on the Texas coast by primary metals producers. As a result, the area no longer imported all of the metals required by its sustained expansion. The primary metals companies that migrated to the area in the modern era thus indirectly facilitated development by lowering the transportation costs of the metals used in many regional industries. At the same time, the construction and operation of their manufacturing plants directly contributed to the growth of the regional economy by creating the foundation for the future expansion of a major new industry. Since the nineteenth century, primary metals production has been a symbol for economic maturity, and its growth in and around the refining region marked a significant step toward a mature, diversified regional economy.[21]

Another major new industry, the construction and operation of the National Aeronautics and Space Administration's Manned Spacecraft Center, firmly linked the region's economy to the expansive scientific-electrical industries that many see as the symbol for economic prosperity in the future. In the early 1960s, NASA selected a thousand acre site southwest of Houston for the construction of its $60 million center. This land had been recently given to Rice University in Houston by the Humble Oil and Refining Company, and Rice in turn donated it to NASA. During the 1960s, the expansion of the Manned Spacecraft Center created almost 5,000 jobs while encouraging the growth of numerous closely related industries. Although the economic stimulus generated by the space complex waned with the decline of the manned space program itself, the NASA facilities permanently altered the regional economy by attracting a variety of high technology industries and companies to the Houston area.[22]

The "traditional" secondary industries in the region—that is, cotton, timber, and grains—also continued to expand in the modern era. Houston and Galveston continued to serve as processing and shipping centers for the newer cotton producing sections in northern Texas and Oklahoma. The spread of the cotton culture was accompanied and facilitated by the ability of the coastal cities to perform their traditional financial and commercial functions over broadening geographical areas. While modern changes in the methods of growing cotton displaced much of its former work force, the level of production did not decline, and shipments from Houston and Galveston continued to challenge those from the Gulf Coast's traditional cotton port, New Orleans.[23] Although the level of total exports from the region decreased after World War II, cotton remained a significant secondary industry. Its effects were concentrated in Houston and Galveston, however, and it had little influence on other coastal cities.

Conversely, the business of exporting timber accrued largely to the Beaumont-Port Arthur area. As with cotton, a departure in the scale of operations marked the evolution of the lumber industry in the modern era. Several large, national paper and pulp producers built major plants in the coastal region. The Champion Paper and Fibre Company's Pasadena plant continued to expand after its construction in 1937, and several other large, modern manufacturing plants for the processing of timber also grew near Beaumont. The Kirby Lumber Company used control of vast acreage in East Texas to become a giant among state and national lumber companies, and its plant twenty miles north of Beaumont expanded into one of the largest, most modern lumber manufacturing facilities in the South in the 1950s. In the same period, the Houston Oil Company joined with a new partner, Time, Inc., to build a large paper plant, the East Texas Pulp and Paper Company, north of Beaumont. These three coastal plants illustrated a trend that was evident throughout the East Texas lumber industry in the post-World War II era—the movement toward large manufacturing plants owned by nationally-based companies. As such companies invested substantial sums in permanent coastal facilities for the manufacture of finished goods from lumber cut in the interior, they strengthened the traditional economic ties between the upper Texas Gulf Coast and East Texas, where reforestation promised a continuing supply of timber. Lumber thus remained a significant regional industry, giving employment to substantial numbers of workers in the counties just north of Beaumont and in the timber region as a whole.[24]

The regional economy also felt the influence of the continued expansion of rice growing in the modern era. This activity was concentrated in the low-lying coastal areas near Beaumont and Lake Charles, Louisiana, and throughout the century the coastal region remained one of the major rice producing sections in the nation. Other grains, however, came to overshadow rice, at least in terms of total shipments from regional ports. The potential profits to be reaped from a Gulf Coast port for the export of midwestern grain was the lure that had originally convinced Arthur Stilwell to build his railroad from Kansas City to Port Arthur in the 1890s. Although oil temporarily postponed the utilization of this trade route, grain shipments had increased gradually after the initial oil boom, and by the 1950s and 1960s such shipments became an important export for the ports of Houston, Beaumont, Galveston, and Corpus Christi.[25]

THE ASCENDANCY OF THE COASTAL REGION

Urban and industrial growth in the diversifying coastal region in the modern era climaxed a century long shift in the distribution of population and industry in eastern Texas and western Louisiana. Table 4.9

Table 4.9. Value Added by Manufactures in Each Region: 1900–1963 (in thousands of 1926 dollars)

	1900	% of Total	1929	% of Total	1947	% of Total	1963	% of Total
Coast	18,016	53	177,133	78	512,527	87	1,630,502	91
Cotton	6,228	18	8,092	4	13,950	2	45,829	3
Mixed	3,075	9	7,678	3	14,760	3	26,510	1
Timber	6,865	20	35,416	16	44,763	8	88,267	5
Total	34,184	100	228,318	100	586,001	100	1,791,108	100
Total -Coast	16,168	47	51,185	22	73,474	13	160,606	9

Source: U.S. Census of Manufacturers 1900, 1929, 1947, and 1963.

provides a rough estimate of the long-run changes in sectional manufacturing activity. The indicators used—valued added by manufacturing and number of manufacturing workers—suggest that a substantial, sustained expansion of the section's total output was accompanied by the increasing concentration of sectional manufacturing within the coastal region. Similar trends characterized the related process of urbanization. As is evident in Table 4.10, the urban centers on the coast were the locus of sectional population growth. In the modern era, as in most of the century, there were no large cities in the timber and cotton regions, and the coastal cities, as well as a circle of cities (Austin, Waco, Dallas, Fort Worth) surrounding these regions, grew at the expense of the interior.

Statistics on the expansion of Houston help place regional growth in a national context. Table 4.11 reveals the city's rapid ascent through the urban hierarchy. In the seventy-five years since 1900, the nation's eighty-fifth ranked city catapulted into the fifth highest ranking, surpassing the population of Detroit and Baltimore since the 1970 census. The pace of its growth was most rapid in the 1920s and the 1940s, when Houston and other southwestern cities recorded one of the fastest rates of urban growth in the country. Thus Houston's sudden "emergence" in the modern era as one of the nation's major urban centers was in reality the culmination of a century of growth. Whether Houston or the entire coastal region is studied, the conclusion is the same: the rapid expansion of total population, urban population, and industrial output which have

Table 4.10. Population In The Coastal Cities*, 1900-1970

	Total Urban Population	As Percent of Coast	As Percent of Section	Percent Change from Previous Decade
1900	103,264	64	11	--
1910	161,060	64	14	56
1920	267,504	69	20	66
1930	477,628	73	29	79
1940	579,256	66	31	21
1950	876,768	67	39	51
1960	1,280,242	66	45	46
1970	1,570,356	63	45	23

Source: Census of Population, 1900-1970.

*Includes Houston, Galveston, Beaumont, Port Arthur, Orange, Lake Charles.

characterized the modern era is a continuation of, not a departure from, earlier trends.

This sustained expansion of the coastal region has been in part at the expense of surrounding sections. Table 4.12 suggests the extent of the relative decline of the cotton, timber, and mixed regions. The cotton region grew most slowly throughout the century as changes in the methods of production and the westward shift of cotton growing produced a steady decline in its rate of growth. But even this region—though lagging behind the rest of the section—showed a slight absolute population increase in each decade before 1930. In the modern era, however, each decade brought an absolute decrease in population. The decline of the major cotton counties was even more rapid than is evident from the charts, since several counties included for convenience in the cotton region were not major producers and did not, therefore, suffer the precipitous loss of population which took place in the major cotton

Table 4.11. Houston in the Urban Hierarchy, 1900–1960

City	1960 Rank	1960 Population	1930 Rank	1930 Population	1900 Rank	1900 Population
New York	1	7,781,984	1	6,930,446	1	3,437,202
Chicago	2	3,550,404	2	3,376,438	2	1,698,575
Los Angeles	3	2,479,015	5	1,238,048	36	102,479
Philadelphia	4	2,002,512	3	1,905,961	3	1,293,697
Detroit	5	1,607,144	4	1,568,662	13	285,704
Baltimore	6	939,024	8	804,874	6	508,957
Houston	7	938,219	26	292,352	85	44,633
Cleveland	8	876,050	6	900,429	7	381,768
Washington	9	763,956	14	486,869	15	278,718
Saint Louis	10	750,026	7	821,960	4	575,238

growing areas. The fate of those areas is illustrated by the history of Fayette County, located between Houston and Austin. In 1900 it was the third largest cotton producing county in the state, and its population of 36,500 was only 8,000 less than that of the city of Houston. By 1930 it had slipped to 30,700; by 1970, to only 17,700. Such a pattern recurred throughout the cotton region and throughout much of a larger cotton producing area which extended along the blackland prairie into northern Texas. The timber region and the mixed region, which contained both cotton and timber production, were more prosperous, although their growth fell far short of that of the coast. The steady, increasingly rapid decline in their percentage of total sectional population in the modern era continued a trend which had begun even earlier.

These general trends are useful in understanding the forces that accounted for the sectional realignment of population. The urban, industrial centers on the coast grew steadily throughout the century, exerting an attraction to potential migrants that was particularly strong during the rapid coastal expansion of the 1920s and the 1940s. In the cotton region, this pull was matched by the constant push of declining rural opportunity, a force that was especially strong after the onset of the Great Depression and the mechanization of cotton production tech-

Table 4.12. Regional and Sectional Population Growth: 1940–1970

	1940	1950	1960	1970	% Change 1940-1970
Coast	871,134	1,305,716	1,928,499	2,501,775	187
Cotton	486,018	420,538	409,528	415,324	-15
Mixed	202,424	182,002	174,259	202,155	0
Timber	330,953	312,295	314,987	382,108	15
Total	1,890,529	2,220,551	2,827,273	3,501,362	85
Total -Coast	1,019,395	914,835	898,771	999,587	-2

See pages 20 and 21 above for an explanation of these divisions.

niques after World War II.[26] The mixed region experienced a somewhat slower pace of out-migration until the 1940s and 1950s, when the growing attraction of the coastal cities increased the outward flow of population. The timber region, by contrast, exerted a temporary pull of its own in the early decades of the century. Its power to attract migrants never approached that of the coastal region, however, and after about 1920 the opportunities on the growing Gulf Coast began to attract migration from the timber region. In the section as a whole, the coastal boom during World War II, following as it did a decade of depression that was most severe in the interior regions, brought a major realignment of population. The decade from 1940–1950 witnessed the most widespread shifts of the century, thus supplying much of the work force that sustained the development of the coastal region in the modern era. (See Table 4.13.)

A CENTURY OF REGIONAL GROWTH REVIEWED

The most powerful economic magnet attracting migration from the interior to the coast throughout the century was the industrial activity spawned by the petroleum industry. Petroleum refining, the manufacture of oil tools and supplies, and the production of petrochemicals set the pace of the industrialization and the urbanization of the coastal region. The direct impact of these manufacturing activities was evident in different parts of the region at different times, and their spread brought significant shifts in population within the region. The initial emergence of Port Arthur and Beaumont as full-fledged cities after

1901 was the earliest and most obvious instance when the expansion of the petroleum industry drastically altered the pace and geographical distribution of regional growth. Between 1901 and 1931, the refining capacity of Jefferson County increased by almost 300,000 barrels per day; in the same period, the populations of Beaumont and Port Arthur rose from 10,000 to more than 100,000. The rapid expansion of refining along the Houston Ship Channel led a similar population boom in the area between World War I and World War II. In the modern era, Lake Charles, the mid-county area between Beaumont and Port Arthur, Orange, and Brazosport all experienced rapid growth in response to the expansion of petroleum refining or petrochemical production. As the primary manufacturing concerns in the region, these major industries shaped much of the pace, timing, and location of twentieth century growth on the upper Texas Gulf Coast.

Petroleum-related manufacturing was thus a "growth pole" that generated other industrial activities while itself continuing to grow. The construction of more and more refineries and petrochemical plants throughout the century reinforced the impetus for regional growth originally provided by the discovery of the Spindletop oil field. Each new plant represented still another growth pole implanted in the regional economy. As shown above, the cumulative result was the growth and spread of a massive refining complex, one that was large enough to sustain substantial regional development even in the absence of the growth of other related industries.

As is expected of a successful growth pole, however, refining did encourage the expansion of other industries. Through the linkages between the coastal refineries and other regional and national industries flowed a strong stimulus for economic expansion and diversification. To estimate the magnitude of such stimuli, economists have developed input-output or interindustry analysis, which follows the flow of transactions among sectors of an economy in order to estimate the impact of the purchases of each sector on all other sectors.[27] Input-output analysis thus allows economists to predict the final effect that one transaction will have as it works its way through the remainder of the economy, and, in so doing, it allows the historian to approximate the impact that the growth of a particular industry has exerted on other economic activities in the past. Several interindustry studies in the modern era have included parts of the upper Texas Gulf Coast, and their findings suggest the magnitude of the overall economic impact of the growth of refining and petrochemical production in that region.

Table 4.13. Each Region as Percentage of Sectional Population: 1900-1970

	Coast	Cotton	Mixed	Timber	Total -Coast
1900	18	47	15	20	82
1910	23	40	14	24	77
1920	29	35	13	23	71
1930	40	30	11	19	60
1940	46	26	11	18	54
1950	59	19	8	14	41
1960	68	14	6	11	32
1970	71	12	6	11	29

The state of Texas sponsored one such study of the entire state economy in 1970. Because it is a state-wide analysis, this study does not isolate the influence of the refining industry on the coastal region. It does, however, suggest the regional impact of petrochemical production, since 88 percent of all petrochemicals produced in Texas in 1970 came from plants located along the coast. The value of shipments from this industry (in Texas) totaled about 4.4 billion dollars. The direct and indirect spending induced by the production of petrochemicals in Texas in 1970 was, however, even higher, with estimates ranging between 10.5 and 14.6 billion dollars. Respending effects thus made petrochemicals a very potent agent of development within the remainder of the economy.[28]

As earlier input-output study of the Port Arthur-Beaumont economy arrived at similar conclusions about petroleum refining in that area. This analysis revealed that in 1955 alone petroleum refineries in Jefferson County expended the following amounts for "inputs" from the surrounding area: $36 million for crude oil, $27 million for chemicals and rubber, $20 million for fabricated metals products, $13 million for utilities and communications services, $87 million for transportation, $58 million for construction, $120 million for households (which included salaries, wages and profits), and $37 million for taxes. Portions of these dollars were then spent and respent in the economy, thus encouraging an expansion of output throughout the regional economy. The final "income multiplier" calculated for petroleum refining in the Sabine-Neches area as of 1955 was 1.917, which suggests that each one dollar change in income from petroleum refining ultimately induced an almost two dollar increase in total area income.[29] Although such income multipliers cannot be assumed to remain constant over time, it is safe to conclude that the expansion of petroleum refining in the coastal region exerted similarly powerful multiplier effects throughout the twentieth century. Through these ties to the regional economy, petroleum refining and petrochemical production successfully performed the long-run function of growth poles by stimulating the expansion of other industries, thereby encouraging the diversification of the region's industrial base.

This process of diversification was most successful in and around Houston, which came to dominate the regional and sectional economies as it matured as a city (see Table 4.14). Before the discovery of oil, Houston had been the regional transportation center, and its strong economic connection to the cotton and timber region and its administrative and financial leadership had given it substantial and lasting advantages over cities like Port Arthur. After 1901, Houston benefited much more from the growth of all branches of the petroleum industry than did its rivals to the east. Even before the opening of its ship channel in 1914, Houston had become a center for oil production activities in the Southwest. Especially significant in its early industrial growth was the expansion of oil machinery production, which became important to the city immediately after the Spindletop discovery and continued to prosper throughout the century. Beginning in the 1920s, however, the growth of refining exerted an even stronger economic influence. During that decade, the initial expansion of refining along the ship channel spurred a 110 percent increase in Houston's population. From about 1920 through World War II, the large refineries in the Houston area were a primary source of the dynamism that allowed the city to far out-strip its old rival Galveston and to pose a serious challenge to New Orleans' historical dominance of Gulf Coast shipping. During and after the war, the area economy became much more diversified, as the growth of large-scale petrochemical and chemical production along the coast to the south and east of Houston gave the area a valuable new petroleum-related industry. In this period significant import substitution also began, as large producers of primary metals were attracted to the area by the needs of the oil industry and by the more general growth in demand

Table 4.14. The Growing Dominance of Harris County (Houston) in the Region and the Section, 1870–1970

	Population of Harris County	Harris as % of Coastal Region	Harris as % of Section
1870	17,375	34	5
1880	27,985	36	5
1890	37,249	36	4
1900	63,786	39	7
1910	115,693	46	10
1920	186,667	48	14
1930	359,328	55	22
1940	528,961	61	28
1950	806,701	62	36
1960	1,243,158	64	44
1970	1,741,912	70	50

Source: U.S. Census of Population, 1870–1970.

that resulted from urbanization and industrialization. In the 1950s and 1960s, the city's traditional role as the regional center for financial, administrative, and cultural services took on an added economic importance owing to the increasing demand for these services throughout the maturing regional economy. In addition, Houston's growing importance in the national economy attracted a variety of national firms, especially in oil-related industries, and the result was a dramatic increase in financial and administrative activities in the city, a change best symbolized by the rapidly expanding skyline of Houston in the 1960s and 1970s. The city had thus become the region's metropolis, and the expansion of this role and the continuing growth of Houston's industrial base made it both the national symbol for and the prime beneficiary of the economic maturity of the coastal region.

NOTES

1. U.S. Department of Commerce, "U.S. Petroleum Refining: War and Postwar" (Washington, D.C., 1947), p. 92.

2. American Petroleum Institute, *Petroleum Facts and Figures,* 1971 edition (Washington, D.C., 1971), p. 307.

3. *Ibid.,* pp. 314–24.

4. *Ibid.,* p. 135.

5. *Ibid.,* pp. 578–79.

6. *Ibid.,* pp. 572–77.

7. The capacity and total production for Standard is taken from Ralph and Muriel Hidy, *Pioneering in Big Business, 1882–1911* (New York, 1955), pp. 410–17, 187, 284.

8. Gulf coast capacities were taken from *Oil Investor's Journal,* March 22, 1971, pp. 94–123. For "free world" demand in 1969, see *Petroleum Fact and Figures,* p. 574.

8. Throughout the modern era, Gulf Coast refineries continued to handle large amounts of crude from the southwestern United States. But as more and more crude was imported from foreign fields, refineries in other locations—such as the Virgin Islands and the eastern United States—were better situated to receive this crude, refine it, and ship it to East Coast markets. For an overview in changes in the location of refining, see *Petroleum Facts and Figures,* pp. 560–71.

9. John Frey and H. Chandler Ide (eds.), *A History of the Petroleum Administration for War, 1941–1945* (Washington, D.C., 1946), deals with the World War II experience in government-business coordination. For a good summary of the performance of the oil industry during World War II, see Harold Williamson, et al., *The American Petroleum Industry: 1899-1959: The Age of Energy* (Evanston, Illinois, 1963), pp. 747–94.

10. Report to the Special Senate Committee Investigating Petroleum Resources by a Committee Appointed by Officers of the API and Others, "A Study Relating to the Disposition of Government Owned Refineries and Related Facilities" (Washington, D.C., 1945), pp. 5–50.

11. As discussed earlier, Lake Charles had begun to grow in 1920s as a rice processing center, but it expanded rapidly in the modern era under the impetus provided by refining and petroleum production.

12. All statistics on refining capacity are taken from the U.S. Department of Interior, Bureau of Mines yearly survey, "Petroleum Refineries in the United States and Puerto Rico."

13. For statistics on the decline in refinery employment, see Edward J. Williams, "The Impact of Technology on Employment in the Petroleum Refining Industry in Texas, 1947-1966," unpublished doctoral dissertation, University of Texas—Austin, 1971. The impact of technological change on the refinery work force is discussed below, on pages 177-183.

14. For a discussion of the various products grouped under the generic term "petrochemicals," see Arthur Browstein (ed.), *U.S. Petrochemicals, Technologies, Markets, and Economics* (Tulsa, 1972). For a listing of all the various petrochemical producers on the upper Texas Gulf Coast, see Houston Chamber of Commerce, '74–'75 Houston Gulf Coast Chemical Directory (Houston, 1974).

15. *Oil and Gas Journal,* January 28, 1959, p. F-38.

16. Norman Whitehorn, *Economic Analysis of the Petrochemical Industry in Texas* (College Station, Texas, 1973), pp. 1–30.

17. *Ibid.,* p. 7.

18. The Census of Manfactures' Standard Industrial Classification (SIC) 28, Chemicals and Allied Products, contains chemicals as well as petrochemicals. But on the Gulf Coast petrochemicals are far and away the largest component of SIC-28. In fact, advances in

petrochemistry are rapidly increasing the use of hydrocarbons in a variety of products traditionally supplied by the chemical industry.

19. Most of this feedstock is taken from crude oil and natural gas in the refineries, and the Gulf Coast has traditionally been the primary supplier of such feedstock for the national petrochemical industry.

20. For a history of the growth of petrochemicals in the Sabine-Neches area, see Joseph Clark, *The Texas Gulf Coast: Its History and Development* (4 vols.) (New York, 1955), v. 2, pp. 243-55.

21. Joseph Clark, *The Texas Gulf Coast,* v. 2, pp. 151-61. See also, Houston Chamber of Commerce, *Houston Economic Handbook, 1955—Metals Industries* (Houston, 1955), foreword.

22. David McComb, *Houston, The Bayou City* (Austin, 1969) , pp. 203-5.

23. For export figures, see Department of the Army, Corps of Engineers, *The Port of Houston* (Washington, D.C., 1936), p. 115; *The Port of Galveston* (Washington, D.C., 1936), p. 97 in Port Series No. 6. For New Orleans, see *The Port of New Orleans* (Washington, D.C., 1932), p. 190 in Port Series No. 5.

24. *Beaumont Enterprise,* October 10, 1954, Section VII, pp. 1-5.

25. See, for example, the export statistics in Department of the Army, Corps of Engineers, *Waterborne Commerce of the United States,* Calendar Year 1958, Part II (Vicksburg, Miss., 1959), pp. 74-120.

26. James H. Street, *The New Revolution in the Cotton Economy: Mechanization and its Consequences* (Chapel, Hill, N.C., 1957).

27. Input-output analysis records the flow of goods between industries. After a descriptive table of transactions between sectors is compiled, mathematical operations performed on this table yields an inverse of the matrix table. This table shows the direct and indirect requirements of each industry after all purchases have worked their way through the economic system studied. (In this respect they are similar to Keynesian multiplier effects.) See Raymond Goldsmith (ed.), *Input-Output Analysis: An Appraisal* (Princeton, N.J.: National Bureau of Economic Research, 1955).

28. Herbert Grubb, *The Structure of the Texas Economy,* 2 vols. (Austin, 1973).

29. C.D. Kirksey, *Interindustry Study of the Sabine-Neches Area of Texas* (Austin, 1959), pp. 48, 49, 121, 122.

Part II

Oil and Institutional Change

Introduction

THE CONTOURS OF INSTITUTIONAL CHANGE

A defining characteristic of twentieth century economic growth in America has been the central role played by large-scale organizations. "Big Business," "Big Labor," and "Big Government" have come to exercise increasing power over the process of growth and the distribution of the resulting costs and benefits. As the expansion of the petroleum industry tied the upper Texas Gulf Coast to a dynamic sector of the national economy, large international corporations, nationally based labor unions, and a variety of government agencies at both the state and federal levels took an ever greater role in the region's economy. These organizations formed the institutional framework within which were made the most important decisions affecting regional development. Part II examines the histories of some of the crucial institutions that comprised this evolving framework and then suggests how the relationships among these institutions changed over the course of the twentieth century.

Because they directly controlled much of the regional economy, the vertically integrated oil companies which owned the coastal refineries were the most powerful institutions in the region. Decisions reached by the individual oil companies—often in headquarters far removed from the region—reverberated throughout the region's economy, dictating the pace and direction of much of its development. For most of the twentieth century, these center firms expanded their operations on the Gulf Coast, thus providing a strong impetus for regional growth. Before the 1930s, their decisions were rarely challenged by others in the region, and even after the rise of independent labor unions and the emergence of other important industries on the Gulf Coast after World War II, the vertically integrated oil companies remained central to the institutional framework that governed the region's evolution.

The needs of the growing oil companies greatly affected the development of other important economic and political institutions. Numerous corporations engaged in oil-related activities migrated to the region to supply goods to the large oil companies or to take advantage of opportunities created by the growth of refining. The evolution of a different type of economic institution, industrial labor unions, was even more closely tied to the growth of the oil industry. The giant coastal refineries created the preconditions for the rise of such organizations by attracting large numbers of industrial workers to the region, and the gradual emergence of an independent, nationally affiliated oil workers' union then brought important shifts in the distribution of economic and political power. The needs of the oil companies and the general problems created by rapid industrialization also shaped the evolution of several existing political institutions while calling forth the creation of others. Of course, the structure and function of these political agencies differed from those of labor unions, which in turn differed from those of the corporations in the oil-related complex of industries. Nonetheless, the histories of these diverse institutions exhibited certain similar patterns due to the influence of the major petroleum companies active in the region.

An understanding of these similarities requires an overview of the changing division of labor among regional institutions. The general periodization used in Part I is again useful. In the early years (1901 to about 1916), the largest oil companies' dominance of the booming petroleum industry made them the most powerful organizations in the region. The political and economic institutions inherited from the pre-oil era were not yet prepared to respond to the changes brought by the discovery of oil, and they only gradually became capable of asserting authority over important aspects of oil-induced growth. In the meantime, the major oil companies had nearly complete responsibility for the control of oil-related functions. They dominated economic activities through their ownership of the main branches of the oil industry and their participation in numerous secondary industries that were later taken over by independent firms. They exercised nearly total authority over those who worked in the refineries. They became primary agents of political change while seeking to avoid political barriers to their continued growth. Although antitrust sentiment was often a temporary check to their expansion, they generally avoided its full effects. In short, the formative years were marked by the oil companies' direct control over most important aspects of the process of development.

This imbalance began to change amid the unusual economic circumstances brought by World War I. The middle period (1916 to about 1941) was not marked by smooth, continuous change, and, in certain ways, the 1920s and the 1930s seem to belong to two different eras. They are united, however, by a common and significant theme—the gradual emergence of a variety of secondary institutions with the capacity to coordinate, to augment, and occasionally even to replace the power of the primary regional organizations, the largest oil companies. This process of institutional differentiation took varied forms, as many functions earlier performed or closely supervised by the oil companies were taken over by specialized organizations outside of their direct control. Independent firms which grew to supply goods and services needed in the refining process, labor unions which asserted authority over the cost of labor and the conditions of employment, the American Petroleum Institute which took the lead in oil pollution control efforts, and political agencies which established control over a range of oil-related issues all took over functions that had previously been carried out in large part by the oil companies. Part of this shift reflected the effects of the Great Depression, which brought a general movement away from the cooperative institutions of the 1920s and toward the public sector. But the political activism of the New Deal era was itself a part of a broader trend that had been apparent in the 1920s—the search for institutional arrangements capable of dealing with problems that the individual firms, no matter how large or powerful, had proven incapable of adjusting.

In the modern period (1941 to the 1970s), the movement toward institutional differentiation climaxed in a period of institutional diversification, as several economic and political institutions matured into organizations capable of challenging aspects of the oil companies' traditional power in the region. In the coastal region, the economic symbols for this new era were the large petrochemical plants and the primary metals producers. In labor relations, independent organizations firmly established control over most aspects of working conditions in the refineries. Their growing power brought a measure of diversity to regional politics; through its political action committees, the Oil Chemical and Atomic Workers (AFL-CIO) became a political force capable of exercising a significant influence on regional, state, and national elections—and on public policy. In the era of diversification, public agencies came to exert increasing power in the regulation of such oil-related problems as pollution. Although the large corporations were certainly not helpless in the post-World War II period, they were no longer as powerful relative to the remainder of the society as they had been earlier. Their sustained growth had helped to create other institutions outside of their direct control that were increasingly capable of independent actions.

Part II examines the histories of some of the most important of these institutions, with special emphasis on their changing relationships to the vertically integrated oil companies that remained at the center of the regional economy. Chapter V focuses on the evolution of economic institutions by describing the development of the increasingly diverse oil-related industrial complex that remained the heart of the region's economy. Chapter VI provides an overview of the changing institutional arrangements for dealing with the oil industry's work force. Chapter VII is a study of some of the most important government agencies that asserted authority over many oil-related functions. The final chapter supplies a much more detailed examination of one of these functions, the control of oil-related pollution, that began as a "private" concern and gradually became a "public" concern. It is intended as a case study of the interaction of the institutions discussed in chapters V through VII. Each of these chapters frequently looks outside the region, for the institutions that most affected the region were increasingly national and international concerns.

The decision to follow the evolution of each separate set of institutions over time rather than to discuss the changes in all of the institutions during each general time period followed from a desire to organize the analysis around the familiar historiographical guideposts provided by business history, labor history, political history, and the emerging field of environmental history. These guideposts served as primary points of reference in the attempt to make sense of the complex process of regional development in an increasingly interdependent national economy. They proved instructive in suggesting what to look for in studying the evolution of the different sets of institutions. Unfortunately, the existing historical literature was somewhat less useful in analyzing the connections among the various institutions over time and the more general ties between the development of new institutional arrangements and the process of economic growth. Generalization about this last complex issue is beyond the scope of my history of one particular region, and Part II does not seek to construct a general theory of economic development and institutional change. It attempts the much more modest task of describing the evolution of the oil-related institutions in both the public and private sectors that most influenced the refining region's twentieth century development.

Chapter V

The Oil-Related Complex of Industries

A prominent geographer has observed that the growth of oil-related industries on the Gulf Coast represents a "phenomenon of the same order of magnitude as the traditional iron and steel complexes of the nineteenth century."[1] As iron and steel had done for the Pittsburgh area, this agglomeration of petroleum-linked manufacturing left its distinctive imprint on the development of the region surrounding Houston. At the historical center of this industrial complex stood the large oil companies. Their vertically integrated operations in production, transportation, refining, and marketing gave them direct control over the central branches of the petroleum industry, and backward and forward linkages to other industries further extended their influence.[2] The economic activity spawned by the needs of the oil companies became increasingly diverse as the oil industry grew, as some of the supply industries expanded into other markets, and as the opportunities afforded by oil-induced development attracted a variety of other center firms to the region. The size and complexity of the resulting collection of petroleum-linked economic activity defies comprehensive analysis. I have chosen instead to examine several of the most important industries involved: oil-related construction, the production and fabrication of metals for such construction, the supply of goods and services used in the refining process, the transportation industry, and the production of petrochemicals. By focusing on the changing relationships of these industries to the petroleum industry, I have sought to present a historical, dynamic account of the growth of the oil-related industrial complex which dominated the regional economy.

AN OVERVIEW OF THE EVOLUTION OF THE OIL-RELATED INDUSTRIAL COMPLEX

The process of vertical integration influenced both the size and the institutional structure of the oil-related industrial complex. It allowed a scale and stability of operations that led to the construction of refineries large enough to serve national markets. As shown in Part I, many of these giant refineries were built on the upper Texas Gulf Coast, and they helped attract other economic activities to the region. Some of these—most notably the production of crude oil and its transportation to the refineries—were undertaken by the companies that owned the refineries. Others were performed by a variety of independent firms which grew to supply the often specialized needs of the oil companies. The lines between the two were not clearly drawn. In fact, the economic tasks performed by the oil companies changed over time, and an overview of this change is important in understanding the evolution of the oil-related industrial complex.

Immediately after the discovery of the Spindletop field, companies such as Gulf Oil and Texaco faced a difficult struggle for survival in competition with other local firms and with the Standard Oil combination. At the same time, they had to secure many essential goods and services not readily available in a regional economy as yet unable to meet the needs of the booming new oil industry. While forced in the short run to perform some such services for themselves and to rely on strong personal contacts in the older eastern fields to secure other services, these emerging firms also sought to fashion organizational structures that would enable them to survive and prosper in the long run. Managers might dream of challenging Standard Oil in eastern markets, but pressing daily concerns rudely interrupted such dreams. How could they obtain much needed fabricated metals from which to construct storage tanks for their crude oil? Where could they find containers, much less tankers, in which to ship their products to market? If their companies were to prosper, these managers had to prevent the hardships imposed by the lack of an established market for oil-related goods and services from stifling their on-going efforts to establish viable operations in the primary branches of the petroleum industry.

In response to these difficult circumstances, the successful companies chose similar strategies: they focused their energies and resources on strengthening their control of production, transportation, refining, and marketing while encouraging the growth of dependable secondary firms that could be relied on to supply numerous goods and services essential to their operations. The choice to concentrate on the basic branches of the petroleum industry was dictated by several considerations. Foremost among these was the desire for predictability. To assure continuity in their operations, the large oil companies took control of the oil from the ground to the markets. Their choices of functions also reflected their judgment of the relative profitability of the various aspects of the oil industry. Especially in the formative period in the histories of Gulf Oil and Texaco, complete vertical integration was further encouraged by the existence of the Standard Oil combination, which served as the accepted model of efficient organization within the industry. Standard was a very powerful competitor capable of stunting the growth of any company that relied too heavily on its well-developed transporting and marketing capabilities. To compete with Standard, the regional companies had to duplicate its structure. They took considerable risks in the process, pouring millions of dollars into the construction of refineries, the search for adequate sources of crude, and the acquisition of pipelines and tankers to transport their products.[3]

Accompanying the process of vertical integration was an opposite, though less pronounced, trend toward a form of disintegration.[4] In the early years, the lack of dependable suppliers of essential goods and services forced the large oil companies to maintain close control over a variety of important secondary manufacturing activities. As the regional oil industry became increasingly well integrated into the national petroleum industry and as this national industry became more complex and diversified, the regional oil companies were able gradually to relinquish direct control over many of these secondary functions. The result might best be labeled "institutional differentiation," since many tasks formerly controlled by the dominant economic institutions in the region, the large oil companies, were taken over by a number of specialized supply firms. Some were nationally active concerns that migrated to the region to take advantage of the expanding demand for their products; others were regional firms that had been involved in other industries before the discovery of oil; still others were established by men whose ties to the oil companies helped alert them to opportunities to build independent supply firms. Some of these opportunities involved taking over functions initially performed by the oil companies; others were created by ongoing changes in the production process of the petroleum industry. In the early years, most of these firms depended exclusively on markets provided by the oil companies, but some later matured into more independent firms serving broader markets. Their evolution represented the second of two significant trends that characterized the growth of the oil-related industrial core in the period before World War II: (1) the concentration of control over the central branches of the petroleum industry in the large vertically integrated oil companies; and (2) the development of strong backward linkages from these companies to a variety of specialized, independent supply firms.

Institutional differentiation—that is, the performance of specialized oil-related functions by an increasing number of firms not directly controlled by the oil companies—in the middle period set the stage for what might best be labeled "institutional diversification" in the modern era. Several changes underlay this move toward an increased number of economic institutions not wholly dependent for markets on the vertically integrated operations of the large oil companies. The first was simply the movement of some of the largest supply firms into markets outside of the oil industry. A second source of both institutional and economic diversification was the growth of several new regional industries, most notably the manufacture of primary metals and the production of petrochemicals. Building on strong forward linkages from the oil industry, petrochemical production added a measure of economic diversity previously missing from the oil-related industrial complex. Along with the primary metals industry, petrochemicals also contributed to the institutional diversity of the regional economy, since the operations of these center firms matched, or at least approximated, those of the oil companies in scale and complexity.

What had begun with the efforts of a few firms to build vertically integrated oil companies while securing essential goods and services in a sparsely developed region thus evolved into an impressive agglomeration of oil-related manufacturing. The largest oil companies retained a central place in this industrial complex. Their refineries remained the largest manufacturing concerns in the region; their pipelines reached out from the Gulf Coast to the major producing fields in the Southwest as well as to the markets on the East Coast; the skyscrapers that housed their administrative headquarters dominated the skyline of Houston. But by the modern era, numerous other industries spawned by backward and forward linkages from these companies had grown into increasingly important sectors of the expanding and diversifying oil-related complex of industries. The historical relationships of such industries to the oil companies, the influence of on-going economic ties among these industries, and the dependence of most of them on the continuing supply of inexpensive fuel from the oil and natural gas companies all created a considerable identity of interest among the various parts of the industrial complex. Nonetheless, this identity of interests did not extend to all matters, and many of the economic institutions in this complex gradually outgrew their historical dependence on the oil companies and became increasingly able to assert powerful, independent voices in the economic and political decisions that determined the course of the region's development.

THE EXPANDING CORE: THE GROWTH OF OIL-RELATED INDUSTRIES

The above overview suggests a general pattern which bears repeating before closer analysis of the evolution of individual firms and industries in the oil-related industrial complex: (1) direct involvement of the large oil companies in most petroleum-related activities in the early years; (2) a gradual differentiation of tasks as other, "backward-linked" firms grew to supply goods and services to the oil companies; and (3) the emergence in the modern era of several important new "forward-linked" industries which used the petroleum industry's outputs instead of supplying it with inputs. The following case studies seek to illustrate and amplify this overview by examining how specific companies were affected by the processes of differentiation and diversification. They are not offered as a comprehensive history of the oil-related complex. They are meant instead to describe the development of some of the most important parts of this complex by tracing the growth of representative companies active in the transportation of petroleum products, petroleum-related construction, the production of metals, the supply of goods and services to the refineries, and the production of petrochemicals. Taken as a whole, these case studies suggest the growing diversity of the oil-related complex while illustrating how its evolution was shaped by the closely related processes of vertical integration and disintegration in the petroleum industry.

Transportation Industries

Whether packaged or shipped in bulk, petroleum products required an extensive and, in part, specialized transportation system to reach markets. In the early years, perhaps the most pressing problem confronting the managers of the growing regional oil companies was the lack of facilities to transport their oil from the fields to their refineries and, finally, to markets in both the Southwest and East. While building a network of pipelines from the major fields to the coastal refineries, the oil companies relied heavily on the best existing means of transportation in the region, the railroads. They gradually acquired oceangoing tankers and barges for use on surrounding rivers and canals and later added fleets of trucks. The oil companies thus owned much of the transportation system that they used, but substantial regional and national concerns grew to build and maintain this system. As with other industries that supplied the oil companies with goods and services, the extent of direct involvement by the companies in these transportation industries varied by function and changed over time.

Because of the size and diversity of the markets they served, the railroads remained more independent from the oil companies than any other industry that transported petroleum products. This also reflected the fact that at the time of the Spindletop discovery, the railroads were much more than the primary means of transportation in the region. As discussed in Chapter I, they were the largest corporations then active on the Gulf Coast, and their repair shops in Houston and Beaumont were among the most important manufacturing concerns in those cities. While connecting local industries to national markets, they also facilitated the flow of investment by attracting to the region the capital of both those who owned the railroads and their friends. Their activities thus largely determined the pace and direction of pre-oil development. In short, in 1900 the railroads were what the oil companies would later become: the center of the most important financial and industrial complex in the region.

As such, they were keenly interested in the opportunities offered by the Spindletop discovery. In pursuing this interest, they quickly developed close ties to the regional oil industry. The most basic of these ties was the service they provided to the oil companies by transporting essential goods to the oil fields and then carrying the oil produced to the coastal refineries or to markets in the interior. Indeed, the conversion of most southwestern railroads from coal-burning to oil-burning locomotives meant that the railroads were themselves one of the largest markets available to the regional oil companies.[5] Such interdependence was strengthened by strong financial bonds between the two industries. The heavy investment in Texaco by the most prominent owner of the Kansas City Southern, John Gates, was not unusual, as many of those connected with the railroads which served the new oil fields sought to get in on the ground floor of the young oil companies established there. The impetus for railroad investments in oil came from the oil companies as well. Texaco's dealings with the International and Great Northern Railroad suggest the identity of interest that developed between buyer and seller. In pursuing a contract to supply the I&GN's fuel, Arnold Schlaet, who was in charge of marketing at Texaco, approached the railroad's president, George Gould, with the idea of buying into the Texas Company. In offering Gould stock at a discount. Schlaet reminded him of the close connections between their companies:

> I desire to make the distinct statement that this is not merely a proposition to sell you a block of stock, but that I believe we are offering you a good deal. As before stated, we consider it good business for the Company, as well, to have more of the stock in the hands of present and prospective customers.[6]

Gould accepted the offer, subscribing for $100,000 in Texaco stock. In a subsequent letter to the president of Texaco, Gould acknowledged the mutual advantages of the deal by thanking him for his "kind offer that you and your friends in the section south of Pittsburgh will help us all they can when we get ready to extend there."[7] The interdependence of the railroads and the large Gulf Coast oil companies made them natural allies in developing the regional oil industry.

The alliance could become quite strained, however, when it came time to divide up the profits generated by this development. The basic points of contention for the oil companies were the quality of service and the shipping rates for petroleum; the railroads were most concerned about the price they paid for fuel oil. In the disputes that inevitably arose, the railroads, which were larger and more firmly established than were most of the oil companies, held several other significant advantages. They could control the development of new oil fields by extending or failing to extend new lines. They could bargain for lower fuel prices by shifting their considerable purchases to competing oil companies or by threatening to switch back to coal if oil prices rose too high. Their ultimate threat was that they would enter the petroleum industry and aggressively compete with existing oil companies. Several railroads moved beyond idle threats by actually establishing their own oil production companies. The following description by the president of Texaco of a meeting he had with the purchasing agent of the Southern Pacific Railroad suggests that the road's production company, Rio Bravo, was used to influence Texaco's asking price for oil:

> He told me . . . he was opposed to the Rio Bravo extending their business, either in pipe lines or otherwise, if it could be avoided; that they had enough to do, to attend to railroad business without extending their oil interest, so long as they could procure supplies at a satisfactory price, all of which I made mental note of, and of course was pleased to hear.[8]

Of course, the oil companies were not confined to making mental notes of all the high cards played by the railroads in this high stakes game for the profits generated by the petroleum industry. In fact, their ability to build pipelines, thereby eliminating the need for the railroads' services, gave them the trump card, and they played it several times. In 1905, for example, Texaco constructed a pipeline from the Humble field near Houston to the main line of the I&GN, thus ending the Southern Pacific's monopoly on traffic to and from that field and provoking a bitter controversy that finally reached the courts. On another occasion, Security Oil Company apparently forced the Southern Pacific to lower its freight rates merely by threatening to build a pipeline from Louisiana to Beaumont.[9] Amid such threats and counterthreats, the two industries for a time maintained an uneasy coexistence.

Motivated by this uneasiness, the oil companies ended their dependence on the railroads for transportation as soon as their finances permitted by constructing an extensive pipeline network. More gradually, the growth in demand for petroleum products removed another source of dependence by making the railroads' purchases of fuel oil less vital to the oil companies' economic health. Even the financial ties between the two industries weakened as the sustained growth of the oil companies gave them access to other sources of capital. Although the railroads' strong ties to the emerging oil industry thus proved temporary, they were nonetheless important. The dominant regional industry of the late nineteenth century had provided crucial assistance and transportation services to its twentieth century successor before the petroleum industry gradually built its own transportation system as a part of its vertically integrated operations.

Pipelines played a pivotal role in this transition by giving the oil companies control over the overland shipment of their product. The construction of an extensive pipeline network thus took a high priority in the early years, and numerous lines were quickly laid from the coastal fields to the refineries. The materials for these lines came from eastern companies such as the National Tube Company, a Pittsburgh-based concern with ties to both Standard Oil and U.S. Steel. Pipe shipped to the region from either Pittsburgh or Birmingham was laid by company crews under the supervision of experienced pipeline specialists hired away from the eastern fields.[10] The first major lines extended to fields far outside the coastal region were the Gulf Oil and Texaco lines to the Glenn Pool, Oklahoma field. Completed in 1906 and 1907, these lines were major investments for their companies, and they represented a strong commitment to the expansion of Gulf Coast refining. In subsequent years, pipelines were built at great expense from the Gulf Coast to all the large fields in West Texas, East Texas, and Louisiana. At the same time, the pipeline network within the region continued to expand, forming the crucial connecting links among the coastal refineries and petrochemical plants and leading oilmen to begin referring to the region as "the spaghetti bowl." Before World War II, oil company crews generally laid these lines using pipe produced outside the region. But in the modern era, the growing demand for pipe helped attract steel mills that produced pipe to the upper Texas Gulf Coast.[11] At the same time, the oil companies came to rely increasingly on independent contractors to lay their pipelines, and several companies specializing in pipeline construction also grew into important parts of this complex. As crucial components of the vertically integrated operations of the large oil companies, the pipeline network thus generated strong backward linkages to other industries both inside and outside the region. Through such linkages, the construction and maintenance of this private transportation system had an economic impact that reached far outside the oil companies that owned the lines.

The same was also true for another portion of the oil industry's private transportation system, the fleets of ships that carried oil from the coastal refineries to markets throughout the world. In the first decade of the century, the growing demand for shipping from both the California and Texas oil industries severely taxed the existing shipping network and the shipbuilding industry which sought to supply new vessels. Such shortages temporarily forced the Gulf Coast oil companies to ship goods to East Coast and European markets in tankers owned by Standard Oil and the Shell Transport Company, which then owned the largest fleets of tankers in the world. The regional oil companies were wary of relying on concerns outside of their control for water transportation, just as they had been reluctant to continue to rely on the railroads for moving oil overland. Their response was predictable and quite expensive: they sought to take control of their transportation needs by acquiring their own ships as soon as possible. This was not easy to do given the scarcity of ships capable of carrying oil. For a time, Gulf Oil, Texaco, and Sun Oil scoured the nation searching for ships that could be converted into oil tankers, and when they found suitable vessels, they often ended up bidding against each other for the same ships. As soon as their finances permitted, each company sought to remedy this situation by placing orders for new tankers specially designed for the oil trade.[12]

These orders introduced a new era in the development of the shipping capacities of the Gulf Coast oil companies, an era characterized by the acquisition of fleets of oil tankers from large shipyards in the eastern United States. Gulf Oil led the way with the purchase of the S. S. *J.M.*

Guffey from the New York Shipbuilding Company's Camden, New Jersey, yard, and it subsequently continued to secure many of its large vessels from that company. In 1908 Texaco purchased its first new oil tanker, the S. S. *Texas,* from the Newport News Shipbuilding & Dry Dock Company of Newport News, Virginia, and it later acquired tankers from other large East Coast shipyards.[13] Texaco decided to take its involvement in the shipbuilding industry one step further in 1912 by organizing the Texas Steamship Company. In the next decade, the expanding needs of Texaco and the unusual demand for ships during World War I spurred the growth of this company's shipyard at Bath, Maine, where thirty-five vessels of assorted types were built in the period from 1916 to 1921.[14] Texaco then discontinued its shipbuilding operations and joined the remainder of the oil companies in the region by returning to its earlier policy of acquiring ships from established eastern shipyards. After World War II the escalating price of shipbuilding in the United States led many of these oil companies to turn to foreign builders, and by the 1970s many of the largest and newest tankers were built by foreign companies and then leased to the oil companies while registered under "flags of convenience" in nations such as Liberia.[15] Although these tankers were an important extension of the transportation system which led to and away from the Gulf Coast refineries, the economic impact of their construction was thus concentrated first in the eastern United States and then later in other parts of the world. The oil-related industrial complex on the Gulf Coast centered around the large refineries; other regions throughout the world benefited more directly from other activities, such as shipbuilding, which were also essential to the oil industry. The glue holding together the various regional economies that performed such oil-related functions was the vertically integrated operations of the largest oil companies.

Although large oil tankers were not constructed in the region, the oil companies did acquire most of their so-called "small ships," including tugs and barges, from regional shipyards. Such boats carried petroleum products on the rivers and canals of the Gulf Coast, and their construction and repair encouraged the growth of several large shipyards in the region. The size and number of these concerns, as well as their products, were largely determined by the needs of the oil industry; as one commentator observed, "area shipyards tend to be, in fact, huge metal-fabrication shops which derive more income from serving the needs of Gulf coast refineries and oil producers than from marine construction and repair."[16] Many of these yards specialized in such oil-related marine equipment as offshore drilling rigs. Through the operations of these yards, some of the economic benefits generated by the expansion of the large oil companies' private water transportation system remained in the regional economy.

Construction

In contrast to tanker construction, almost all of the economic benefits generated by refinery construction accrued to the region. Due to the sustained growth of refining on the Gulf Coast, such benefits were substantial throughout most of the century. The individual oil companies were originally closely involved in aspects of construction, but they gradually came to rely on nationally active petroleum construction specialists for the planning, the materials, and the manpower needed to expand their refineries.

In the boom years after Spindletop, the major regional companies looked to established firms from the eastern oil fields to provide them with the materials needed for the construction of their first coastal refineries. Faced with a scarcity of both materials and skilled manpower, the managers of Texaco, Gulf Oil, and Security all turned to construction companies with close financial and personal ties to their respective firms. From the goods provided by such concerns, the work force of the oil companies constructed their own refineries. Gulf Oil and Security (Magnolia) chose Riter-Conley Manufacturing Company, a Pittsburgh-based iron and steel construction firm with strong ties to the Mellon family, the dominant financial interest in Gulf.[17] Texaco relied on even closer contacts. Before migrating to the Texas oil fields, Texaco's president, Joseph S. Cullinan, had helped establish the Petroleum Iron Works of Pennsylvania (PIW), an iron and steel construction company that originally catered to eastern petroleum markets.[18] In building oil storage tanks and then the Texaco Port Arthur refinery, Cullinan gave most of his orders to the PIW, in which he retained a large stock share and a measure of administrative influence.[19] Such direct ties had obvious advantages for Cullinan as an individual investor, but as the president of Texaco, Cullinan also found them very useful in securing prompt, dependable service in a period marked by shortages of essential products due to the demands placed on existing supply and construction companies by rapid refinery growth throughout the nation.[20]

As refinery construction became more complex, however, the major oil companies increasingly abandoned such bonds of familiarity with general metal works firms and turned instead to reliance on a growing group of national specialists in refinery construction. In general, these companies had diversified into petroleum and petrochemical construction from related pursuits. Most were originally eastern firms, but almost all established strong ties to the Gulf Coast in response to its sustained growth in refining capacity. The Lummus Company began in Boston, primarily as a designer and builder for the industrial alcohol industry. The business was purchased in 1930 by a company specializing in combustion engineering. As one division of this diversified company, it became a worldwide designer and builder of petroleum refineries, chemical plants, and other processing facilities.[21] M.W. Kellogg Co. began in 1905 as a manufacturer of forge-welded, special steel products. It gradually became a petroleum refinery specialist, handling several large projects for Magnolia's Beaumont works in the 1920s and 1930s. Dresser Industries, Inc., moved into refinery construction from a related interest in oil field supplies and tools, and Brown and Root did so from more general construction work. The Fluor Corporation charted a path that was different only in a geographical sense; it entered the business in California and moved eastward to the Gulf Coast.[22] These firms, as well as several other major refinery contractors and designers active on the Gulf Coast, matured into national or international concerns with interests in various types of industrial construction. Some became diversified oil, petrochemical, and natural gas construction divisions in larger firms like Dresser Industries, which also supplied other services and goods to the same industries. No matter how such firms entered refinery construction, however, once they were in the business, the fast-growing refining complex on the Texas-Louisiana Gulf Coast became one of their largest markets.

These construction concerns merit special attention because of the scale of their regional operations. They did not simply build one plant and then leave. The sustained growth of refining and the related expansion of petrochemical processing plants meant that these construction firms were active in the region throughout the century. Given the rapidly changing technology of the refining industry, they were also periodically called upon to replace obsolete equipment. Their role as builders and rebuilders of refineries gave them a continuing role in the regional economy as suppliers of services and as employers. In the early years, oil company personnel did much of the construction work with materials supplied by these companies. During the booms in refinery construction in the early 1920s and the late 1930s, construction workers often built the plants and then secured production and maintenance jobs in them.[23] In the post–World War II era, the contractors generally provided their own labor, hiring some workers locally and others with special skills from outside the Gulf Coast. In addition to skilled blue-collar workers, these companies brought specialized engineering, technical, and managerial experts into the region. Those that were most active on the Gulf Coast eventually built administrative offices and manufacturing plants in the region, usually in Houston.[24]

By the modern era, the vertically integrated oil companies relied almost entirely for construction of refineries and for other oil-related construction on the expertise, experience, and specially trained personnel of these companies. The early ties between the oil companies and many of these construction specialists remained evident in the expanding number of petroleum-related services which they offered, but the largest of them also cultivated a variety of worldwide markets outside of the petroleum industry. While remaining an important and active segment of the oil-related industrial complex on the Gulf Coast, companies such as Brown and Root also linked the regional economy to a construction-related industrial complex of diversified, internationally active construction specialists and the various secondary industries that supported them. The largest of the construction specialists thus exhibited a pattern that also characterized the production of petrochemicals and primary metals: their sustained growth gradually made them center firms that were not solely dependent on the markets provided by the regional or even the national petroleum industry.

Metal Industries

Petroleum-related construction created a great demand for different types of metals, and the industries that grew to meet this demand made up an essential part of the evolving oil-linked industrial complex. The production of fabricated metals for the construction of refineries and oil storage tanks, the manufacture of oil tools for drilling wells, the operation of foundries to cast such metals, and finally, the production of primary metals for use in these foundries all became major regional industries. The companies that undertook these activities were the largest regional manufacturing concerns other than the refineries. As these companies grew, they generally became more and more tightly integrated into the national metals industry, and their close connections to the major producers of metals in the nation ultimately helped to attract to the region the large manufacturing plants of several of these large corporations.

From the earliest years after Spindletop, the oil tool and supply companies that furnished the equipment needed to drill for oil were perhaps the most active secondary firms in the region. Houston's rail connection to the East made it the central supply point for the Spindletop field, and many of the oil field supply firms that serviced that field established headquarters at Houston. A series of discoveries at Sour Lake (1901), Batson (1901), Humble (1904), and Goose Creek (1908) further encouraged the expansion of these Houston-based firms by moving the center of coastal production toward the city. Even after the discovery of vast new fields in Oklahoma, West Texas, and East Texas, oil production along the coast still helped support a substantial oil tools industry. By 1941 there were 185 producing fields within a 150-mile radius of Houston, and more than 400 oil tool manufacturers and distributors in the city supplied these surrounding fields. The largest of these were national supply firms which used Houston as their base of operations from which to provide services throughout the southwestern oil industry.[25]

A comparison of the evolution of two such companies that became important regional manufacturers, Hughes Tool and Ideco, reveals two very different patterns of growth. Hughes Tool Company was founded in 1909 by two men who had been involved in the regional oil industry from its inception: Howard Hughes and Walter Sharp, who was also associated with a production company controlled by Texaco. These men saw a golden opportunity in producing high quality drilling bits for the booming southwestern oil industry. Drawing on their extensive first-hand knowledge of drilling, they developed a revolutionary new bit and built a giant, Houston-based company around its manufacture and sale. Hughes Tool Company quickly grew into the largest producer of drilling bits in the world, and, primarily because of its expansion, the Houston area economy came to supply most of the bits used throughout the international petroleum industry.[26] A small Beaumont-based firm followed a somewhat more circuitous path in becoming a large international concern. The S. and H. Supply Company was a locally owned company founded in response to Spindletop. But through a series of purchases by larger concerns, this successful local supplier became a part of an international company. First, it was absorbed by Boykin Machinery and Supply, a regional concern. Then, in 1929, the International Derrick and Equipment Company (Ideco) purchased Boykin. Finally, in 1944, the company became a part of the worldwide, diversified Dresser Industries, Inc. During this time, Ideco became an important national producer of rotary oil drilling rigs and a large regional employer.[27]

A number of local companies that sold fabricated metals to the growing coastal refineries took a path similar to that of S. and H. Supply, prospering as small, regionally active firms until being absorbed by national corporations. The largest early suppliers of fabricated metals, however, migrated from eastern fields and generally had close ties to individual oil companies. Petroleum Iron Works of Pennsylvania and Riter-Conley of Pittsburgh served as both the original construction company and the prime metal supplier for Texaco and Gulf Oil, respectively. When Joseph Cullinan left Texaco in 1913, PIW, in effect, went with him, becoming an important part of the diversified American Republics Corporation. PIW and a second American Republics' subsidiary, Pennsylvania Shipyards, built extensive new manufacturing facilities in Beaumont in the 1920s. But after the parent company faltered, these subsidiaries were sold to U. S. Steel and Bethlehem Shipyards.[28] John Dollinger, Inc. was more successful; it prospered after it moved to Beaumont in 1906 and acquired markets throughout the region.[29] Cameron Iron Works of Houston outgrew local markets and became an international, diversified company.[30] But the experience of Beaumont Iron Works was perhaps most typical of those regional companies that survived after the Spindletop boom subsided. A builder of railroad cars before Spindletop, this firm became a metals supplier to the oil industry and to the railroads. In 1945 it was acquired by American Locomotive Company, a national firm active in both fields.[31]

About this time, regional metals production in general underwent a significant change as a result of the entry of a group of subsidiaries of the leading national producers of iron and steel products. U. S. Steel's American Iron and Bridge Division (1940), Bethlehem Steel (1947), Chicago Bridge and Iron (1941), and American Smelting and Refining Company (1943) all established local plants that became major regional producers of fabricated metals. The demand for fabricated metals in the Gulf Coast petroleum market first attracted the attention of these large, diversified metals firms, and they entered the area either by acquiring an existing concern or by building their own regional manufacturing plants. These companies often resembled in scale and organization the regional metals supply firms that had successfully expanded to supply broader and more varied markets. Such fabricated metals producers became increasingly independent of individual oil companies and the oil industry in general. Petroleum specialists who produced fabricated metals had supplied the needs of the regional oil industry in its early history, but in the post-World War II era, this market was taken over by center firms that were among the nation's leading manufacturers and distributors of all types of metal products.

The large metal companies did not restrict their regional activities to fabrication. Until the 1940s, primary metals demanded by the growing refining region came primarily from Pittsburgh and from Birmingham, Alabama. But the continued expansion of iron and steel use in the area did not go unnoticed. In 1940, the oil and gas industries in the Houston-Port Arthur areas used almost 650,000 tons of iron and steel. After a market survey had pointed out this large demand, Sheffield Steel Corporation, a subsidiary of Armco, constructed a large steel mill near Houston. Completed in 1942, this mill had grown to an annual capacity of about one million tons by 1955. By that year, seven other producers of primary metals had followed Armco to the Texas Gulf Coast. A total investment of a quarter of a billion dollars, aggregate employment of almost 10,000 workers, and combined annual sales of more than $350 million (all figures as of 1955) made the manufacturing plants of these eight corporations a potent force in the postwar economy of the Gulf Coast.[32]

These concerns were not as closely tied to the oil companies by their histories or by the final demand for their products as were other parts of the oil-related industrial complex. Nonetheless, the expanding petroleum-linked market for metals, the availability of inexpensive oil and natural gas for fuel, and the general construction activity induced by the growth of refining and petrochemical production had helped attract them to the Gulf Coast. They were, of course, also closely tied to numerous regional firms that sold finished metal products to the petroleum industry. The prime significance in the regional economy of these primary metals producers was not, however, the extent to which they were connected to the oil-related complex, but rather the fact that they tied the region into a segment of the national economy other than the oil industry. Companies like Armco and Alcoa were established national concerns with options to locate their manufacturing plants throughout the nation or the world, and their movement to the Gulf Coast well symbolized the fact that the coastal economy was becoming more thoroughly integrated into an increasingly interdependent national economy.

Refinery Supply Industries

The daily operations of the large refineries required a variety of goods and services, and several important regional industries grew in part to supply these needs. Two of these merit special attention due to their size and to the unusual evolutionary paths they took. One, the sulphur industry, produced a good that was once vital in the refining process. The other, the packaging industry, supplied containers for the shipment of refined petroleum products. Both were at one time closely linked to markets provided by the large oil companies, but as these markets were altered by changes in the production process, the ties between the sulphur industry and the petroleum industry weakened while those between the packagers and the oil companies became stronger.

In the first decades after Spindletop, the expansive petroleum industry greatly influenced the growth of sulphur manufacturing on the Gulf Coast. In a frenzy of exploratory activity after the first gushers had come in on a salt dome near Beaumont, drillers sank wells on almost every salt dome in the region. In the process, they discovered sulphur deposits that were later developed—although not by the oil companies. In 1905 Texaco, one of the most active early oil producers in the region, passed up the opportunity to develop the sulphur-rich Bryan Mound, south of Houston. Nearly a decade later, Freeport Sulphur Company became a major national producer of sulphur by exploiting this large deposit. In the early 1920s, Texaco discovered another giant source at Hoskins Mound (also south of Houston). In spite of an opportunity to enter a potentially highly profitable venture in a closely related industry, Texaco was at that time not prepared to diversify out of the petroleum industry, and it made an arrangement with Freeport Sulphur to develop the deposit.[33] The reasons for this decision are not clear, but apparently Texaco felt that it should focus its energies on expanding its oil production rather than accepting the added risk—and opportunity—presented by its somewhat inconvenient discovery of what was then one of the largest sulphur deposits in the world.

Texaco refrained from developing its find in spite of the fact that the refining process required substantial amounts of sulphuric acid. Indeed, the refining industry ranked second only to the fertilizer industry in its demand for sulphur in the 1910s and 1920s.[34] The sulphur companies on the Gulf Coast thus had an assured, expanding market in the growing coastal refining complex, and in these years several regional companies prospered in part by supplying the needs of the oil companies. The advent of first thermal and then catalytic cracking after World War I gradually reduced the demand for sulphur in the refineries, since these new distillation techniques did not use as much sulphuric acid as had traditional "straight-run" distillation. By this time, however, two regional sulphur companies, Freeport Sulphur and Texas Gulf Sulphur, were firmly established in other markets, and in subsequent decades they were among the largest firms in the national and international sulphur industry. The relationship between these companies and the oil companies took still another twist in the post–World War II era when the large refiners and natural gas producers began to recover marketable amounts of sulphur from the gases produced in their operations, thus becoming competitors—albeit on a very limited basis—to the sulphur companies which they had once had the opportunity to control.[35]

In the case and package industry, Texaco took an opposite approach from its hands-off relationship to the sulphur industry. As the coastal refineries grew in the early years, they required both wooden and metal containers in which to ship their refined products. Such containers were initially supplied by local manufacturers, several of which grew into substantial concerns in the Port Arthur area. Later, national container producers such as American Can and Continental Can established manufacturing plants along the Houston Ship Channel in part to supply the needs of the coastal refiners. This segment of the container industry thus evolved in a pattern similar to that displayed by the metals industries: national specialists in container production gradually replaced or absorbed local specialists in refinery container production.

Texaco was not, however, satisfied with this arrangement, and it responded by entering the container industry in a way that had a substantial impact on the economy of the Port Arthur area. In 1909 Texaco built a case and package plant adjacent to the shipping terminal at its Port Arthur refinery. The plant originally supplied the refinery with five-gallon cans used primarily in the export trade. As it grew to meet the expanding demand for this and other types of containers, a shook mill for the production of wooden cases was added in 1912. Then in the early 1920s, Texaco invested more than a quarter of a million dollars in building a larger shook mill in Morgan City, a small Louisiana town across the Sabine River from Port Arthur. Until 1935 this mill manufactured most of the company's wooden containers, using lumber cut from land purchased especially for this purpose. A vertically integrated oil company had thus spawned a casing and packaging plant which was itself vertically integrated, at least in that part of its operations that produced wooden containers. The Case and Packaging Division of Texaco moved a step toward independence from its parent company in 1932 when it was reorganized as the Texaco Can Company, a wholly owned subsidiary of Texaco. For several years this new company operated "with a view to drawing trade from concerns other than the parent company," thereby seeking to become a diversified producer of containers with markets inside and outside the petroleum industry. It succeeded so well that it could not keep pace with the growing needs of both Texaco and its new customers. As a result, in 1939 it was reabsorbed by Texaco as the Case and Packaging Division of the Refining Department. As such, it then supplied most of the containers used by Texaco throughout its Gulf Coast operations. Employing as many as a thousand workers to produce more than a hundred different types of containers, Texaco's case and package plant, which was the only such facility owned by a coastal refiner, remained a major industry in the Port Arthur area throughout its unusual history.[36] While most other supply industries "disintegrated" from direct oil company control, this particular plant was integrated, than disintegrated, and finally reintegrated. Its history well illustrates the fact that the line dividing the functions of the vertically integrated oil companies from those of backward-linked supply firms was very lightly drawn and could be erased or even redrawn by the pencil that figured the profits and losses of each particular company.

Petrochemicals and Natural Gas

Forward linkages from the petroleum industry encouraged the growth of the most important new industry to emerge in the region in the modern era, the production of petrochemicals, which used output from the refineries as one of its basic raw materials. A 1951 article in *The Lamp,* a monthly publication of Standard of New Jersey, listed most of the ties between these two industries while summarizing the reasons for the phenomenal postwar growth of the petrochemical industry on the Gulf Coast:

> The Gulf Coast, more than any other comparable area in the land, had what it took to foster petrochemistry. There was a bountiful supply of natural gas and refinery gases for raw materials and economic fuel. There was a reservoir of trained manpower, from skilled labor to Ph.D.'s in the many oil refineries along the crescent. There was access to deep salt water for shipping and certain base chemicals; there was also plenty of fresh water for industrial purposes.[37]

Over the next twenty years, these factors continued to attract petrochemical manufacturers to the area. The crucial role of the refineries in bringing this important new industry to the region was even more clearly shown in a 1973 survey which asked all Texas petrochemical companies the reasons for their choice of location. "Nearness to raw materials" was cited as the number one attraction by forty-seven of the sixty firms questioned, and the large coastal refineries and the natural gas industry were the primary sources for such feedstock.[38] This forward linkage from the oil and natural gas industries thus played a crucial role in drawing the petrochemical industry to the Gulf Coast.

Common ownership of the oil and petrochemical industries further strengthened the ties between them. A 1973 study of the Texas petrochemical industry revealed that 88 percent of the state's capacity was located on the Gulf Coast and that "the six largest firms producing petrochemicals are petroleum companies which have integrated forward to the manufacturing and marketing of petrochemicals since 1940."[39] All of the major regional refiners established chemical plants in the region, usually adjacent to their refineries. Other companies which began as chemical producers, notably DuPont, Monsanto, and Dow, built large plants within the region, as did such synthetic rubber producers as Goodyear, Goodrich-Gulf, and Firestone. In the petrochemical industry which emerged between two existing industries, petroleum and chemicals, the joint ownership of Jefferson Chemical Company near Port Arthur by Texaco and the American Cyanamid Company was not unusual. According to the *Texaco Star,* the new company was "a 50-50 partnership that the two companies felt would benefit by Texaco's knowledge of petroleum technology and supply of raw materials, and American Cyanamid's knowledge of the chemical industry."[40] The resulting market structure resembled that of the post–World War II refining industry: the largest oil companies controlled a large percentage of the petrochemical production, but they shared the market with a growing group of smaller chemical companies and refiners.

Oil, natural gas, and petrochemicals also shared similar sciences and technologies. The type of feedstock used and the final products produced were usually flexible. The problem facing companies in each industry was similar to that traditionally faced by petroleum refiners: what mix of products guaranteed the largest profits. The answer lay in the chemistry and the economics of the interdependent market for refined goods, natural gas, and chemicals; it was continuously altered by a shared quest for further understanding of chemical processes and by shifts in the demand for various, highly competitive products. The complexity of the materials involved encouraged the specialized production of certain substances by particular plants, and the transfer of these substances between plants. Gulf's Port Arthur works, for example, became the world's largest producer of one important petrochemical, ethylene, a product which it then supplied to numerous other Gulf Coast plants.[42] A shared pipeline system facilitated such transfers among companies. They also used similar production techniques, and many of the same suppliers, contractors, and construction specialists. In short, the petroleum, natural gas, and petrochemical industries were closely linked at almost all levels, and in the coastal region, the large oil companies were primary developers and owners of all three.

A common geological history also linked the natural gas and petroleum industries together, for their raw materials were usually found in the same places. Indeed, in the early years, regional oil producers treated the natural concurrence of oil and gas as a nuisance; they often-disposed of the unwanted gas by blowing it off or burning it in spectacular oil field flares. When its value finally was recognized, the largest producers of oil held claims to most known reserves. In 1953, for example, a congressional hearing revealed that three oil companies active on the Gulf Coast—Humble, Phillips, and Texaco—were the nation's largest owners of natural gas reserves. Such companies had an early and continuing role in the production of natural gas.[42]

The fact that these two major regional industries produced energy strengthened the bonds among most of the various segments of the region's oil-related industrial complex. Relatively cheap, abundant, and cleanburning fuel reinforced the backward and forward linkages among these industries while encouraging still other manufacturers to migrate to the Gulf Coast. The oil industry was itself the largest consumer of this fuel. As much as 10 percent of all oil arriving at the coastal refineries in the 1910s and 1920s was burned to produce heat for the refining process. Refineries began switching to natural gas for fuel in the 1920s, and by 1950 refineries and oil and gas field operations used more than one-quarter of the national supply of natural gas.[43] The sulphur industry was also greatly encouraged by the existence of inexpensive fuel, since the cost of fuel required to heat the large quantities of water used in flushing sulphur to the surface was "the final determining economic factor" in the recovery of sulphur at competitive prices.[44] The availability of relatively cheap fuel was also a strong attraction to the primary metals producers who migrated to the Texas Gulf Coast.[45] Indeed, since all manufacturing requires some form of energy, the Gulf Coast's abundance of oil and natural gas, the preferred energy sources of most industries for much of the twentieth century, was a strong attraction to many energy-intensive manufacturers. Especially in the modern era, cheap fuel powered the further expansion of a regional economy already booming due to the growth and diversification of its oil-related industries.

The general importance of the region's supply of inexpensive fuel in attracting industry emphasized once again the crucial developmental role played by the large oil companies. The large corporations that supplied much of this fuel through their operations in the petroleum and natural gas industries expanded the scope and the scale of economic activity on the Gulf Coast. The center firms in oil were much larger than previous area firms; they responded to national, not regional, forces; they had access to similarly large financial institutions; they implemented the most modern technology. Their own impact was reinforced by their close ties to other growing regional industries, as well as many expansive center firms in other parts of the national economy, and such linkages encouraged the emergence of significant secondary and tertiary activity in the regions.

The size and diversity of the resulting oil-related industrial complex made the upper Texas Gulf Coast unique among regions with economies closely tied to the petroleum industry. As iron and steel had done for the Pittsburgh area in the late nineteenth century, or automobile production had done for Detroit in the mid-twentieth century, this agglomeration of petroleum-linked manufacturing gave the coastal region surrounding Houston its identity within the national economy.

The giant coastal refineries were the heart of this industrial complex. Through economic ties from the refineries to other branches of the vertically integrated oil companies and to backward- and forward-linked industries flowed a powerful impetus for sustained growth. As the region prospered, some of the firms originally linked closely to the oil industry gradually became more independent, thus contributing a measure of economic and institutional diversity to the regional economy. Even so, the imprint of oil's historical dominance remains clear. The economy of the upper Texas Gulf Coast as yet contains an increasingly diverse oil-related industrial complex, not a complex of diverse industries. Although the latter is perhaps a worthy goal for a maturing region, the former will continue to serve the region well as long as oil, natural gas, and petrochemicals continue to play a vital role in the modern economy.

NOTES

1. Peter Odell, *An Economic Geography of Oil* (London, 1963), pp. 189–190.

2. A backward linkage is the tie between a primary industry and another industry which supplies a good or service utilized in its production process. A forward linkage is the tie between the primary industry and an industry which uses its output as an input in another activity. See Albert O. Hirschman, *The Strategy of Economic Development* (New Haven, 1958), p. 100.

3. For information on the process of vertical integration in the oil industry, see John McLean and Robert Haigh, *The Growth of Integrated Oil Companies* (Boston, 1954).

4. For a perceptive and suggestive theoretical discussion of this process, see George Stigler, "The Division of Labor is Limited by the Extent of the Market," *Journal of Political Economy* 59 (June 1951), pp. 185–193.

5. For an overview of the shifting demand from coal to oil in this period, see Harold Williamson et al., *The Age of Energy* (Evanston, Ill., 1963), pp. 167–205; also, National Industrial Conference Board, Inc., *Oil Conservation and Fuel Oil Supply* (New York, 1930);

and Sam Schurr and Bruce Netschert, *Energy in the American Economy, 1850–1975* (Baltimore, 1960).

6. Arnold Schlaet to George Gould, letter dated September 1, 1905, Drawer 1-A, Papers of Joseph S. Cullinan, Gulf Coast Historical Association Archives, Houston Metropolitan Research Center, Houston Public Library.

7. George Gould to Joseph Cullinan, letter dated October 23, 1905, Drawer 1-A, Papers of Joseph S. Cullinan.

8. Joseph Cullinan to Arnold Schlaet, letter dated November 1, 1904, *Texas Company Archives,* vol. 36, p. 119, in the Texas Company Archives, White Plains, New York.

9. *Oil Investors' Journal,* September 3, 1905, Arnold Schlaet to Joseph Cullinan, letter dated October 10, 1904, *Texas Company Archives,* vol. 35, p. 170.

10. *Beaumont Enterprise,* February 2, 1901. *Oil Investors' Journal,* January 18, 1907.

11. Joseph Clark, *The Texas Gulf Coast: Its History and Development,* 4 vols. (New York, 1955), vol. 2, pp. 152 and 164.

12. *Port Arthur Herald,* August 3, 1901, p. 1. Information about shipping conditions in the early years is found in *History of The Texas Company's Marine Department,* published in 1952 by Texaco's Marine Department and available in the Texas Company Archives. See, also, Joseph Cullinan to Arnold Schlaet, letter dated September 24, 1904, *Texas Company Archives,* vol. 34, pp. 199–200.

13. *National Oil Reporter,* February 22, 1902; *Oil Investors' Journal,* March 5, 1907, May 5, 1908, and November 6, 1908.

14. *History of The Texas Company's Marine Department,* pp. 11–16.

15. Such arrangements have begun to come under scrutiny in the 1970s due to the rash of oil spills involving tankers registered under the Liberian flag. Little information, however, is yet available about the history of this practice. See Sheldon Novik, "Dunking Liability at Sea," *Environment,* January/February 1977, pp. 25–27 and 44; also, "How the Arcane World of the Argo Merchant, Other Tankers Works," *Wall Street Journal,* January 18, 1977, pp. 1 and 22.

16. C. D. Kirksey, *Interindustry Study of the Sabine-Neches Area,* p. 80. See, also, Joseph Clark, *The Texas Gulf Coast,* vol. 2, pp. 167–172.

17. For Gulf's construction, see *Port Arthur Herald,* February 22, 1902; for Security, Mobil Oil Company publications staff, *History of the Refining Department of the Magnolia Petroleum Company* (Beaumont, n.d.), p. 22. The Mellon ties are reflected in the inclusion of R. B. Mellon on the board of directors. See *Poor's Manual of Industrials, 1918* (New York, 1918), p. 311.

18. Joseph Cullinan to W. C. Baldwin, letter dated April 12, 1906, in Drawer A-2, Papers of Joseph Stephen Cullinan.

19. Memo dated September 30, 1905, in Drawer A-1, Papers of Joseph S. Cullinan.

20. A. W. Krouse to Cullinan, letter dated March 1, 1902, in Drawer A-1, Papers of Joseph Stephen Cullinan.

21. The Lummus Company, *Meet the Heat Exchanger Division* (1954), folder in the American Petroleum Institute Library. Also *Moody's Industrial Manual, 1970,* pp. 2580–2581, entry under Combustion Engineering, Inc.

22. For M. W. Kellogg, see *Poor's Industrial Manual, 1928,* p. 2961; *Moody's Industrial Manual, 1954,* p. 2607; *Beaumont Enterprise,* April 12, 1926, p. 1; *Beaumont Enterprise,* December 12, 1935; for Dresser Industries, see *Moody's Industrial Manual, 1954,* p. 192; for Brown and Root, see *Moody's Industrial Manual, 1971,* pp. 2186–2187; and for the Fluor Corporation, see *Moody's Industrial Manual, 1971,* p. 902.

23. *Beaumont Enterprise,* October 18, 1936, p. 1; January 6, 1938, p. 5.

24. As of 1975 the following major refinery construction firms had large administrative or productive facilities in the Houston area: Dresser Industries; Brown and Root; M. W. Kellogg Co.; Fluor Engineering & Construction; and Fish Engineering and Construction.

25. *Beaumont Enterprise,* October 5, 1941, p. 2F. See, also, Bureau of Business Research, University of Texas at Austin, *Directory of Texas Manufacturers, 1947* (Austin, 1947), pp. 79-93; and Joseph Clark, *The Texas Gulf Coast,* vol. 2, pp. 161-172. Clark's detailed study includes information about almost every major company active on the Gulf Coast of Texas, and I relied heavily on it throughout these case studies.

26. For a sketch of the early history of this firm, see *Oil Weekly,* February 2, 1918, pp. 23-24. See, also, Joseph Clark, *The Texas Gulf Coast,* vol. 2, pp. 162-163.

27. *Beaumont Enterprise,* August 3, 1926, Section A, p. 6; and October 110, 1954, Section III, p. 6.

28. PIW is listed as a division of U.S. Steel in the *Directory of Texas Manufacturers, 1947,* p. 18, while in earlier reports it had not been. C. D. Kirksey, *An Interindustry Study of the Sabine-Neches Area of Texas* (Austin, 1959), pp. 88-89, reports the sale of the Pennsylvania Shipyards to Bethlehem in 1947.

29. *Beaumont Enterprise,* September 10, 1954.

30. *Moody's Industrial Manual, 1973,* vol. 1, pp. 223-224.

31. *Moody's Industrial Manual, 1954,* p. 1872.

32. Houston Chamber of Commerce, "Fabricated Metals Industry," *Houston Economic Handbook,* 1955, pp. 1-27; Joseph L. Clark, *The Texas Gulf Coast,* vol.2, pp. 151-156; and R. H. Startzell, "Texas Steel," *Texas Journal of Science* 2 (December 30, 1950), pp. 451-461. For a history of one of these companies, see Don Whitehead, *The Dow Story* (New York, 1968). In addition to Sheffield, these companies were Aluminum Company of America, which built a plant near Corpus Christi in 1949; American Smelting and Refining Company, Houston, 1943; Dow Chemical Company, Freeport, 1940; Lead Products Company, Houston, 1946; Reynolds Metals Company, Corpus Christi, 1952; Tenn-Tex Alloy and Chemical Corporation, Houston, 1951; Tin Processing Company, Texas City, 1941.

33. William Haynes, *The Stone that Burns: The Story of the American Sulphur Industry* (New York, 1942), pp. 118 and 183.

34. Werner Duecker and James West (eds.), *The Manufactur of Sulfuric Acid* (New York, 1959), pp. 1-8.

35. Jared E. Hazleton, *The Economics of the Sulphur Industry* (Baltimore, 1970), pp. 27-33.

36. *Beaumont Enterprise,* August 30, 1936, p. 4-D. Also, "The Texaco Can Company," in *Refining Department History Compiled During July 1926,* on file at the Texas Company Archives, which contains a variety of sources of information about the Case and Package Division and the shook mill in Morgan City. See file 7720 in the Texas Company Archives for several mimeographed sheets of information about the shook mill and the timber acquired for use in it. See, also, *Texaco Star,* February 1924, pp. 4-9 and February 1923, pp. 6-11. For general information about the packaging operations, see*Texaco Star,* May 1922, p. 9 and *Oil and Gas Journal,* January 17, 1935.

37. *The Lamp* 33 (June 1951), pp. 3-6.

38. Norman C. Whitehorn, *Economic Analysis of the Petrochemical Industry in Texas* (College Station, Texas, 1973), p. 9.

39. Ibid., p. 17.

40. "Petrochemicals: A New Pace for Petroleum," *Texaco Star,* Winter 1960-1961, pp. 5-9. See, also, Houston Chamber of Commerce, *'74-'75 Houston-Gulf Coast Chemical Directory* (Houston, 1975).

41. C. D. Kirksey, *Interindustry Study of the Sabine-Neches Area,* p. 54.
442. Edward Neuner, *The Natural Gas Industry* (Norman, Oklahoma, 1960), pp. 7–8; J. C. Youngberg, *Natural Gas: America's Fastest Growing Industry* (San Francisco, 1970).
43 James R. Stockton, *Economics of Natural Gas in Texas* (Austin, 1952), p. 292. For specific statistics on fuel use in refining, see U. S. Bureau of Mines, "Reports of Investigations," numbers 2964, 3028, 3145, 3220, 3281, 3332, 3367, 3485, 3554, 3607, and 3703.
44. William Haynes, *The Stone That Burns,* p. 68.
45. Joseph Clark, *The Texas Gulf Coast,* vol. 2, p. 151.

Chapter VI

Labor Institutions in the Coastal Refineries

The evolution of independent labor unions in the Gulf Coast refineries followed a pattern similar to that displayed by many of the region's other economic institutions. As with the supply of other inputs into the refining process, the oil companies' direct control over their supply of labor in the early period gradually declined with the rise of independent labor organizations capable of exerting increasing control over wages and conditions of labor. These oil workers' unions were among the first large-scale organizations of industrial workers in the coastal region and the entire southwestern United States. Their growth marked a significant shift in the institutional framework that governed regional development, for the oil workers' unions grew to exert a powerful voice, largely independent of the oil companies, in the economic and political affairs of the region. These unions were concentrated in the refineries and in the petrochemical plants, and they were much less active in the other branches of the vertically integrated operations of the oil companies. The history of their growth and of their changing relationship to the large corporations that owned the refineries falls into three general periods: (1) the nonunion years from 1901 to 1915; (2) the era of organization, 1915 to 1945; and (3) the consolidation of power and then the erosion of this power by rapid technological change after 1945.

THE FORMATIVE YEARS, 1901–1915: THE CORPORATION AS LABOR ORGANIZATION

In the early years after Spindletop, those who owned the refineries exercised direct control over almost all the conditions that affected those who worked in them. The corporations were clearly the primary "labor" institution affecting these workers. During the first wave of refinery construction, the individual plant manager was a labor organizer on a broad scale, since he had to put together an entire work force in a region with no previous refining industry. After recruiting sufficient skilled and unskilled labor, oil company officials then determined wage rates, the length of the work day, seniority practices, and most other work conditions. Initially, management's actions were restrained only by the necessity to attract and maintain an adequate labor supply. Such total sway was short-lived, however, because a process akin to the "disintegration" of the oil companies' direct control over supply industries was also at work in the "labor supply industry." The growth of independent labor organizations was not, of course, exactly the same as that of suppliers of cans or fabricated metals. The owners' fears of the consequences of autonomous workers' organizations made them fight to keep control over their labor force. In spite of this often intense opposition, however, the twentieth century history of the region's refining workers has been characterized by the gradual, often halting, evolution of increasingly independent labor unions.

The labor institutions that emerged were shaped by three primary groups: those who owned the plants, those who worked in them, and those who staffed the various government organizations that intervened during prolonged disputes. The owners desired a stable, efficient work force. In the first three decades of the century, this goal was operationally defined as owner control over an open shop, since independent unions were viewed as a disruptive, destabilizing force. The majority of the refinery workers wanted primarily more money, better working conditions, and job security.[1] The goal of the government in disputes between labor and management was usually the amelioration of the immediate crisis at hand. The evolution of labor institutions was greatly influenced by the changing relative power of each of the three groups to impose its own goals on the others.

In the formative period this power rested almost completely with the large oil companies. A variety of factors strengthened their positions, preventing the emergence of labor institutions capable of competing with the corporation for control over working conditions. Some of these influences were distinctly regional, and some reflected national trends. The following analysis of the most important barriers to unionization focuses on the period before 1915, but many of the conditions described continued to slow the growth of independent labor organizations throughout the country.

The chaos created by the Spindletop boom was not conducive to the formation of workers' organizations. The initial discovery of oil dramatically altered the regional labor market. Rapid migration from surrounding regions and from jobs within the national petroleum industry resulted in a large, and sometimes surplus, supply of workers. Many of these men moved from oil field to oil field in search of money and adventure, and the regional labor market thus was characterized by a level of geographical mobility that was unusually high, even in a very mobile society like the United States. In the early period, and to a lesser degree throughout the century, this source of instability weakened the position of labor unions.

The boom was over by 1906, but the boom mentality remained to hamper efforts at collective action by refinery workers. The scramble for wealth set off by the sudden discovery of oil had been almost a caricature of much of nineteenth century American economic development, and an individualistic, self-help philosophy of "laissez-faire" capitalism thrived in such an atmosphere.[2] It should be remembered that the basic tenet of this Horatio Alger mind-set—that the individual who desired to succeed could do so—was more than an empty promise in this particular time and place. Spindletop further opened up the already fluid social structure of a young, growing region. There were numerous highly visible examples of Alger-type success stories. Indeed, many of the prominent local leaders of Texaco and Gulf Oil had started near the bottom in the oil industry. If nothing else, their experience probably made them less able to accept collectivism, since they understandably felt, "I made it, why can't others?" Those who remained in production jobs in the refineries despite the opportunities for advancement within the oil companies and elsewhere in the booming southwestern society were conditioned by their culture to view their status as a personal choice not to be corrected by collective action.

Cultural attitudes on race presented an even stronger barrier to the emergence of labor organizations. Although the work force had almost none of the ethnic and language factions that proved so troublesome to union organizers in the older industrial regions, the prevailing division between whites and blacks more than compensated for this lack of diversity.[3] Although barred from oil field work in the early period, black workers were extensively employed in refinery labor gangs. As early as 1911, for example, the work force at the Magnolia refinery in Beaumont was about 36 percent black.[4] The attitude of management and white laborers toward this large segment of the refinery force was illustrated by an article on plant activities in Magnolia's monthly news magazine (in

1921): "Mules, niggers, wheelbarrows and a few white men . . . form one of the largest and most essential departments of our refinery."[5] In a society with sanctions against either white and black cooperation or black initiative in forming separate unions, the presence of a large, fixed caste of black laborers served as a powerful barrier to collective action. Since the black labor gang performed the hardest, dirtiest jobs for the least pay, their presence made the lot of white workers both physically and psychologically easier, thereby making them less susceptible to unionization.

Had refinery workers attempted to organize, they could have expected little aid from other area unions. A nucleus of about 1,700 regional workers were organized into numerous small craft unions, but they had little in common with the refinery workers. The railway brotherhoods and brewers in Houston, a substantial group of cotton processing workers in Galveston, and scattered building trades groups throughout the region were the largest of these organizations.[6] None of the existing unions was active in heavy manufacturing industries like refining. They were too small to organize the large refineries and too craft-oriented to include the majority of those who worked in them, and they therefore affected the plants only during a few construction projects.[7] The coastal refineries were equally unaffected by established national unions, since there were no such organizations in the refining industry or in closely related industries.[8] Given this lack of potential sources of aid for organizing refinery workers, the initiative for unionization had to come from the workers themselves.

Powerful legal and public sanctions operated against such initiatives. The area's political attitude toward unions reflected its lack of experience with them. Local public opinion saw unionization as a "foreign" movement that belonged in the older, more heavily industrial and ethnically diverse sections of the nation. Labor organization brought forth images of the confrontations of the 1890s.[9] The region had avoided the widespread violence of that period, and many saw the continued absence of unions as a guarantee against future unrest. The laws of the state embodied a fear of unions that was no doubt encouraged by a lack of experience in dealing with them. Legal sanctions against such activities as picketing were symptomatic of an anti-union bias that continued throughout much of the century and represented an often substantial barrier to the rise of labor institutions independent of the oil companies.[10]

The strongest bar to unionization, however, was the performance of the primary existing labor institutions, the oil companies. When necessary, their power usually could be supplemented by that of the state. In 1902, for example, the supervisor of the construction of Gulf's Port Arthur refinery brought in Texas Rangers to bring back to the refinery a group of laborers who had walked off the job after a wage dispute.[11] Throughout the early period, however, management seldom resorted to the use of such public force; their private strength usually sufficed. Their power rested on control over almost all aspects of the individual worker's lot within the refinery. In the early period, management determined the terms of employment, and labor either accepted them or departed. There is little evidence that the owners of the regional refineries repeatedly abused their power over their work forces. But whether regularly used or not, this power was always there. It became especially evident and effective during labor unrest; a "troublemaker," as defined by the owners, was usually fired. Such arbitrary power was not, however, reserved only for strike-breaking. It was also institutionalized in hiring and firing practices. Under the line-up system, wherein foremen chose or passed over a worker on the basis of their personal opinions of him, the individual worker's job security was in the hands of a low-level and often tyrannical representative of management.[12] If the worker was fortunate enough to be chosen for work, he had little voice in framing the shop rules and procedures which directly affected his job. If he had a grievance, he could confer with his foreman, who held direct power over his ability to earn a livelihood. No collective complaints or suggestions were recognized by management, which justified this stance by citing the "right" of each individual worker to negotiate his own contract free from the dictates of workers' organizations. Such a situation presented few constraints on the treatment of workers by those in charge of the refineries.

In the early period, as throughout the century, however, management relied primarily on the carrot, not the stick. Relatively high wages and good working conditions, not pipe-wielding Pinkerton guards or the six-shooters of the Texas Rangers, were the mainstays of the refinery owners' control over their workers. Although not good by today's standards or even by the prevailing contemporary standards in the more heavily industrialized East, the refinery workers' lot was much better than that of other area workers.

The term "welfare capitalism" did not come into vogue until the 1920s, but its key ingredient, high wages and good working conditions relative to those in other regional industries, was evident in the refining region in the previous decade (see Tables 6.1, 6.2, and 6.3). The best summary of these conditions came from a study published in 1917 by the

Table 6.1. Working Conditions by County and Industry, 1911-1912

Jefferson County	Number firms	Number workers	Average daily wage	Average daily hours	Days in operation per year
Oil company	4	2261	$2.99	9.3	365
Foundry	1	61	2.28	11.75	206
Rice mill	2	60	2.72	9	310
Lumber manufacturing	1	240	2.23	10	263
Harris County					
Oil company	3	411	$3.20	8.3	362
Cotton oil mill	4	610	1.75	12	200
Locomotive & car works	1	56	2.39	10	-
Brewery	1	169	2.77	8	310
Fayette County*					
Cotton compress	1	60	$1,80	10	-
Tyler County**					
Lumber manufacturing	1	331	$2.13	10	267

Source: Second Biennial Report of The Texas State Bureau of Labor Statistics, 1911-1912, Austin, 1912, pp. 74-92.

*Included as representative of a typical cotton county.
**Included as representative of a typical timber county.

Texas Bureau of Labor Statistics: "It will be noted that the average wage paid in the oil industry is considerably higher than in any other, with the exception of a few of the highly skilled trades."[13] Table 6.3 suggests the extent of this differential in the primary refining center of the period, Jefferson County. In 1911, 2,297 refinery workers received an average wage of $2.99 per day, over 31 percent higher than that of the 1,192 workers reported in other occupations, whose average was only $2.29 per day. The manufacturing wages closest to those of the petroleum industry were generally in the foundries, an oil-related industry. Two of the leading pre-oil regional industries, lumbering and cotton oil mills, lagged badly behind the refineries in wages paid. The largest gap, however, was between petroleum firms and nonmanufacturing concerns such as hotels and department stores. Finally, while farming wages were not included in the compilations by the Texas Bureau of Labor Statistics, it seems safe to assume that they also did not compare favorably to refinery wages.[14]

Table 6.2. Average Weekly Wages and Hours for Selected Occupations Paid by the Primary Manufacturers on the Texas Gulf Coast, 1915–1916

	Laborer Weekly Wage	Hours per Wk	Carpenter		Foreman		Machinist		Chemist		Male Stenographer	
			Wage	Hours	Wage	Hours	Wage	Hours	Wage	Hours	Wage	Hours
Beaumont-Port Arthur												
Oil refineries	$13.34	48	$23.13	48	$27.69	48	$22.62	48	$24.67	48	$22.96	48
Saw-planing mill	10.78	60	15.16	60	31.83	60	21.55	60			20.36	60
Electric light plant, street railroads	9.00	70	17.25	60	24.66	70	19.00	60			16.92	60
Houston												
Petroleum products	17.27	51	20.75	50	28.20	48	27.80	48	28.44	56	23.90	45
Cotton oil seed mills	10.75	84					16.79	68	20.16	54	15.00	48
Planing mills	9.26	54	19.72	54	23.35	54					15.83	54
Brewery	10.39	48	19.55	54	34.58	48	22.60	48			23.07	48
Galveston												
Foundries and machine shops	13.50	54			30.62	54	22.97	54				

Source: Fourth Biennial Report of the Texas State Bureau of Labor Statistics, 1915–1916, Austin, Texas; 1917, pp. 30–65.

Table 6.3. Wage Index for Jefferson County, 1911-1912

Type Concern	Number Employed	Average Daily Wage	Index Specific Industry Wage / Average Wage
All firms	3489	2.75	100
Printing and publishing	14	3.28	119
Oil well machinery	12	3.00	109
Oil companies	2297	2.99	109
Steam laundries	172	2.87	104
Boiler works	13	2.75	100
Telephone company	49	2.78	101
Foundry	60	2.72	99
Logging works	22	2.50	91
Electric street railroad	75	2.40	87
Rice mill	61	2.28	83
Lumber manufacture	240	2.23	81
Creosoting plant	44	2.00	73
Lumber transportation	13	1.93	70
Ice factory	38	1.80	65
Feed mill	16	1.80	65
Telegraph company	31	1.57	57
Brick manufacturing	75	1.50	55
Hotel	89	.95	35

Source: Ibid.

Available statistics indicate that, for the most part, the refineries secured the large labor supply required by their rapid early growth with better wages, shorter hours, and steadier employment than their rivals in the regional labor market. In addition to introducing a higher level of wages into the area, the large refineries also were among the earliest regional concerns to shorten the work day from twelve to nine hours and to offer paid vacations of one and two weeks to their employees.[15] Such inducements helped attract and hold a labor force in a region with a limited supply of experienced industrial labor.

The labor market thus came to reflect the "dualism" of a regional economy in which large, nationally active oil companies coexisted with numerous smaller regional concerns. A brief period of cutthroat competition initially kept wages low, but the oil companies that survived quickly raised wages in order to attract and keep the specialized workers they required. In filling certain of these highly skilled positions, these emerging center firms looked to the national petroleum industry. The migration of managers, technical experts, and experienced production workers from the oil fields of the Northeast continued throughout the formative years. The large oil companies retained ties to this specialized pool of labor, which lay beyond the reach of other regional employers. The large refineries bid against each other for such employees, and their competition raised the price of such skills even further.[16] As a result, in both the refinery towns and in Houston, the large oil companies attracted managerial, administrative, legal, and technical talent that was more highly trained than that of most other local businesses. These companies thus held important advantages over other firms in the regional labor market, namely their ability to outbid others for skilled labor in the region while continuing to attract specialists from sources outside the region, including the best educational institutions, the national petroleum industry, and the growing national and worldwide organizations of the individual center firms.[17]

For less skilled labor, the growing oil companies relied more heavily on the immediate area. The shifts in the sectional economy described in Part I greatly expanded the regional labor market, which attracted an influx of migrants from throughout eastern Texas and western Louisiana. The oil companies had first choice of workers from this enlarged pool because they offered the best terms of labor available in the coastal region and the surrounding areas in the interior.

The incentive to pay these high wages arose from conditions characteristic of the center economy. The major refineries brought together a substantial group of industrial workers in several neighboring plants (see Table 6.1), thereby introducing into the region one of the essential preconditions for the emergence of large workers' organizations. The owners' desire to block unionization was among their most important incentives for establishing relatively high wages.[18] Reinforcing this incentive was the economic need to keep the refineries, which grew to a giant scale by supplying national markets, running at near full capacity around the clock. If poor wages yielded an inefficient work force or caused strikes, the owners faced substantial losses.

Several factors also common to the center economy made it possible for the large refiners to pay high wages. Even in the early period, refining was less labor intensive than most other industries, and the increasing pace of technological change after about 1910 further lowered the relative importance of labor costs in the refineries.[19] Even large wage increases did not, therefore, have a significant effect on the total cost of the final product.[20] An emerging oligopolistic market structure also enhanced the refiners' ability to pay. As the most important oil companies grew into center firms, they could more easily pass along increased wage costs to consumers without affecting their competitive positions.

There were, of course, limits to the refinery owners' willingness to increase wages. In these formative years for the regional oil industry (1901–1915), the largest companies of local origin grew steadily toward the status of full-fledged center firms. But they were only beginning to approach competitive parity with other large, national oil companies by 1915. In particular, the superior size and market position of Standard Oil for most of this period meant that regional firms had to hold down costs, including those for labor, in order to compete with Standard. A more important limit on wage rates, however, was the lack of organized pressure from the work force, which generally settled for its relatively favored position in the regional labor market. Although steady growth in the number of refinery workers made them a potentially potent economic force, they lacked the institutional means to define and pursue their group interest.

As a result, the oil companies retained their position as the primary labor organization in the region. As late as 1915 the refinery owners retained almost complete control over their labor force by using both persuasion and coercion to maintain a stable, efficient, nonunion shop. In their dealings with the supply industries discussed in the last chapter, the oil companies usually encouraged "disintegration," which enabled them to rely on secondary firms for many tasks, thus freeing them to focus their energies on the central branches of the industry. In their dealings with their work force, however, they had strong incentives to prevent the disintegration of their direct control. This made it very difficult for workers to establish new institutions outside the corporation in which to assert an independent voice. This was the central task facing them in the next thirty years.

THE ERA OF ORGANIZATION, 1915 TO 1945: A DIVERSITY OF SOLUTIONS

World War I brought significant changes in the existing accommodation between workers and owners. The crucial importance of oil for modern warfare gave a third party, the national government, a strong interest in the continuous operation of the coastal plants, and to facilitate smooth labor relations, the Federal Mediation and Conciliation Service mediated several major labor disputes in the refineries. Under the leadership of craftsmen, refinery workers took advantage of this war-induced change in the balance of power and pressed demands long ignored and even unvoiced. When strikes threatened the production of the major refineries on the Gulf Coast, federal mediators intervened. In the ensuing negotiations, management granted several significant concessions which represented important breaks with labor–management relations in the pre-war years. The agreements that resulted contained the seeds from which a variety of increasingly independent workers' organizations grew in the period between the two world wars.

The first indication of the broad changes to come occurred outside the region, but the war's impact was quickly felt in the coastal plants. The Allies' need for refined products meant that the imperatives of continued production overrode the refinery owners' traditional desire to maintain control of their work force regardless of costs. In July of 1915 workers at the Bayonne, New Jersey, plant of Standard of New Jersey took advantage of the unprecedented vulnerability of their employer. Their attempts to adjust past and present abuses led to a series of violent confrontations that dramatically challenged the existing labor order at Bayonne and in the refining industry as a whole. Against the background of the rising assertiveness of its workers and the tragedy of a strike in which "the city of Bayonne became a battle camp and the refinery a fortress under siege," Standard began to search for new solutions to its labor problems.[21] That company was the traditional trendsetter for refinery wages and working conditions, and its response to the Bayonne strike was closely observed and often imitated in large refineries throughout the nation.[22]

In the coastal region, the initial reaction to the Bayonne strike was the announcement by management of a series of new benefits for refinery workers. The timing of these announcements suggested the impact of the violence at Bayonne. On August 8, 1915, Gulf Oil switched from a twelve-hour day for some production workers and a nine-hour day for the remainder of workers to a uniform eight-hour day. Three days later, the Standard plant at Baton Rouge proclaimed a 10 percent pay raise for all employees. A year later Texaco followed by establishing an eight-hour day, two-week paid vacations, and half pay for National Guardsmen on duty.[23] As in the past, the owners were attempting to avoid possible confrontations by squelching discontent with better pay and improved benefits.

In the new, wartime setting, however, their efforts did not entirely succeed. The refinery workers were not satisfied by benefits alone; they also wanted more say in establishing the shop rules under which they worked. To reinforce their demands, they called the first large-scale refinery strikes in the area's history. The resolutions of disputes at the Port Arthur works of Gulf Oil and the Beaumont refinery of Magnolia merit detailed analysis, since they resulted in significant departures from pre-war labor arrangements.

In the summer of 1916, about 500 skilled workers at Gulf Oil's plant struck for twelve days. Had a federal mediator not intervened the strike probably would have resulted in the walkout of the entire 2,700-man work force.[24] The mediator conducted several conferences between representatives of the workers' unions, the plant manager, and the general manager of Gulf Oil, who traveled to Port Arthur from Pittsburgh to participate in the negotiations. The workers demanded a small wage increase, which they did not receive. But "the principle object of the strike" was the establishment of fixed guidelines governing job security and conditions of labor.[25]

Gulf granted an important concession when it signed with labor representatives an agreement establishing a set of shop rules for the refinery. This accord was hailed by the local labor journal as a "victory for organized labor" and by the mediator as "a new precedent . . . in so far as large oil corporations are concerned."[26] The shop rules included a guarantee of the continuation of the eight-hour day and time-and-a-half for overtime, the recognition of worker-elected shop committees to discuss grievances with management, a pledge of "no firing without proper cause," a thirty-day notice before changes in shop rules, and the establishment of a uniform apprentice system among the crafts.[27] This agreement signaled a significant step toward the establishment of independent workers' organizations. For the first time, Gulf's management had agreed to negotiate with worker representatives. The primary results of these conferences, the shop rules, were the first recognition in principle, if not necessarily in practice, of the workers' right to share in the control of job conditions.

The strike at the Magnolia plant at Beaumont in August 1918 resulted in no such concessions. Its resolution, however, illustrated aspects of the emerging accommodation between worker and owner which were not clearly visible in the Gulf negotiations. The day after 250 boilermakers declared their intention to strike if a 20 percent wage increase was not forthcoming, Magnolia's management announced a 7½ percent raise for all workers. Not satisfied, the boilermakers walked out. Two days later they were joined in an impressive show of solidarity by 700 other craftsmen, production workers, and common laborers. The strike shut down the plant from August 12 to August 24, at which time the second group of strikers were persuaded to return by promises of unspecified future benefits, as well as by the force of public opinion in the community and appeals to their patriotism. Magnolia's plant manager made his position clear in a letter to the federal mediator who had been called in to help resolve the dispute. Citing high wages and good benefits "voluntarily granted" in the past, the manager agreed to allow the strikers to return to work and said he would consider additional wage increases. But he absolutely refused to discuss grievances with union representatives; he stressed that Magnolia had "always maintained and will continue to maintain an open shop."[28] The mediator finally convinced him to agree to a committee grievance plan by citing the opinion of Gulf's Port Arthur superintendent that a similar year-old plan there was "very satisfactory."[29] The official then used his influence with the strikers to convince them to return to work. Unknown to them, the essential elements of the agreement they accepted had been outlined for the mediator in the plant manager's letter. While legitimizing the owner's solution to the workers, the govenment agent also served the interests of the workers by allowing them to negotiate, at least indirectly, with management and by establishing a fragile line for further communications, the grievance committees.

The extent to which management retained control, however, was evident after the boilermakers returned to work on September 19. The strikers had received no additional concessions, only face-saving promises of future improvements by the owners. In the following months the grievance committees were undermined by management's unwillingness to accept the workers' election procedures and by discriminatory treatment aimed at former strikers and union men.[30]

The experience of the workers at the Magnolia refinery suggests a crucial shortcoming of the wartime agreements. Despite a variety of promises, the owners retained complete power over most aspects of the jobs performed by their labor force. On paper, grievance committees and shop rules were definite advances for the workers, but with the withdrawal of government sponsorship of federal unions after the war, the workers were powerless to enforce these agreements. Their weakness reflected the fact that the labor organizations active in World War I had been sustained primarily by the power of the government, not that of the workers. Such unions declined after the war ended and the government lost its direct interest in them.

The disappearance of these "hothouse unions" did not erase several important precedents. For the first time, the refinery workers had experienced the power of organization: they had shut down operations in several of the large coastal refineries. With this new-found power, they had gained a measure of participation in establishing the conditions of labor within the plants. They had seen, however briefly, that management could be forced to deal with the demands of workers' groups. After the war, they would not be content to return to the previous arrangement in which management dictated to the individual worker.

The new accommodation between labor and management which emerged in the twenties embodied many aspects of the changes that had occurred between 1915 and the end of the war. The resulting labor arrangement, nationally known as the American Plan and regionally labeled the Beaumont Plan, rested on management's voluntary recognition of certain rights of labor.[31] This "cooperative" approach often included labor organizations formed and directed by newly hired personnel specialists, whose "scientific management of labor relations" replaced much of the traditional power of the foremen over their work forces.

Many labor historians have dismissed these organizations and the American Plan in general as paternalistic reassertions of the owners' pre-war control. Such criticisms are correct, but far from complete. In the regional refineries, labor relations in the 1920s were similar in important ways to those in the earlier period, but the 1920s also witnessed significant changes. In their haste to get on to the more exciting, more ideologically pleasing events of the 1930s, liberal labor historians, in particular, have tended to describe the worst abuses of the pre-war years, cite the repression of organized labor after World War I, and then skip rapidly over the 1920s to the rise of the CIO. The events of the 1920s, when studied at all, generally have been used as a backdrop for the treatment of the "heroic age of labor" during the New Deal.[32] Such an approach obscures the extent to which labor organizations in the 1920s were forerunners of those in the 1930s. Events in the entire interwar period can, in fact, best be understood as part of a general, long-run trend toward increasing independence of workers' organizations from the control of their employers.

The level of independence possible for refinery workers in the 1920s was severely limited by the growth patterns of the regional oil companies. By then Gulf, Texaco, Humble, Magnolia, and Sinclair had established themselves as equal members of the center economy in oil, and the major refinery of each of these worldwide concerns was in the coastal region. The large work force assembled in the coastal refineries faced severe organizational disadvantages in attempting to bargain collectively with these center firms. To the oil companies, refinery workers were one part of their vast, international, vertically integrated concerns. In dealing with this important, yet relatively small, group of workers, the companies could draw on the political, legal, and financial resources of their entire operations. The organizational structure of labor groups had not kept pace with that of the growing center firms.[33] Workers' organizations in refining were local, not international, and they were restricted to one branch of the industry. Workers in the coastal refineries even had difficulties cooperating with those in neighboring oil fields, since the industrial work force concentrated in the coastal refineries had little in common with the smaller group of less skilled and more highly mobile oil field workers who continually moved to new drilling sites throughout the Southwest.[34] Such a situation created a pronounced imbalance of power between the center oil firms and various local refinery workers' groups, and this basic imbalance was at the heart of the labor-management accommodation of the 1920s.

The oil companies' growth was also accompanied, however, by changes that made them reluctant to abuse their increased power. The rise of new competition to Standard, a change that had begun even before Spindletop, culminated in the 1920s in a relatively stable, oligopolistic market structure in the petroleum industry. The security that this afforded to the major oil companies gave them an even greater ability to pay higher labor costs than in the pre-war years. Their determination to block the rise of "outside" unions remained a strong incentive to improve working conditions, but by the 1920s, another motive became increasingly important. The individual companies and the industry as a whole were becoming more conscious of their public images. Public relations took on an added significance as the large companies attempted to establish in the public's mind the differences between the oligopoly of the "new era" and the monopoly of the not so distant past.[35] The labor violence of the war years had not reflected well on the oil companies' much advertised new "maturity." Peaceful labor relations would make for good public relations. As a result, in the 1920s the plant owners were both willing and able to purchase renewed control over their labor force.

For their part, the workers seemed eager to be bought, but their price was not low. The refinery owners in the region instituted multifaceted reforms that affected many of the conditions of labor. The most comprehensive employee relations plan was probably that instituted at Humble's Baytown plant in the late 1920s. It was borrowed from the parent company, Jersey Standard, which had developed the forerunner of much of the American Plan as applied to refineries after the Bayonne strike.[36] The centerpiece of Humble's employee relations program was the "Joint Conference," which conducted regular meetings between representatives of labor and management. Throughout the region similar efforts were made to establish channels through which to deal with refinery workers' grievances. At the same time, the source of many of these grievances was removed as the newly created personnel offices took control over hiring and firing practices from the individual foremen.

The owners did more than talk to workers' committees twice a year. Most of the large refineries raised wages while also instituting new labor-related programs including the eight-hour day, guaranteed vacations, death and injury benefits, shop rules similar to those established at Gulf Oil's Port Arthur Works in 1917, and safety programs aimed at preventing disabling injuries. To improve the workers' lives outside the refinery, the oil companies took on several new functions, including the provision of free or inexpensive health care for their employees and their families, the construction of low-cost housing near the refineries for their workers, and the establishment of cooperative stores that featured discount prices for all oil company personnel.[37] In addition to these direct benefits, management took an active role in improving the general quality of life in the refining communities. Magnolia, for example, became the leading industrial citizen of Beaumont by building both a 2,500-seat community hall and the area's first radio station. The company's management proudly claimed to be "in the van of humane, progressive, and enlightened employers of labor"[38] Texaco went further in stating that "the secret of quality is contentment of the work force, and Texaco refineries are truly striving to become something more than just a place where oil is refined."[39]

These statements called forth the image of contented workers happily cooperating in surroundings so pleasant as to make the granting of vacations appear almost as a form of punishment. A very different judgment has been passed by most labor historians, who have generally dismissed such plans as public relations smoke screens designed to conceal a harsh, exploitative paternal system.[40] The latter evaluation, though overly severe, correctly identifies the central weakness of the cooperative labor programs of the 1920s: they were paternalistic. Any agreement reached between labor and management was certain to reflect the existing imbalance of power. The workers' lack of power was especially evident in such arrangements as Humble's Joint Conference, which "recognized in principle that the employee had some protection of his job."[41] In practice, the owner retained the ability to define the crucial word "some." In other labor-related matters, management could similarly define the limits of cooperation. The fact that they generally set forth programs that benefited their work force was the basic strength of the accommodation of the 1920s; the fact that workers were essentially powerless if management decided to do otherwise was its basic weakness.

Both the strength and weakness of the situation were reflected in the wage rates paid to refinery workers (see Table 6.4). As in 1911, Jefferson County wage statistics for 1929 reveal a continuing pattern of dualism: the refinery workers received better wages than their counterparts in other local industries. This table also reveals, however, a slight decline in the position of refinery workers relative to that of another group of

Table 6.4. Average Wages and Salaries by Industry, Jefferson County, 1929

	Number Firms	Number Wage Earners	Average Yearly Wage	Wage Index	Number Salaried Workers	Average Yearly Salary	Salary Index
All industry	141	14,386	$1365	100	2037	$2325	100
Printing and publishing	10	55	1504	110	1	2508	108
Planing mill products	3	66	1489	109	21	5416	233
Foundry	11	509	1429	105	60	3741	161
Petroleum refining	5	10,798	1389	102	1360	2262	97
Other manufacturing	47	1,739	1382	101	407	2130	92
Bread	23	171	1364	100	21	1692	73
Ice manufacture	10	120	1360	100	42	2282	98
Beverages	8	42	1297	95	15	1881	81
Coffee	5	14	1243	91	20	1771	76
Lumber and timber products	8	736	973	71	41	2616	112
Rice cleaning	3	86	959	70	27	3168	136

Source: U.S. Census of Manufactures, 1930, vol. 3, p. 516.

oil-related concerns, the foundries. Although refining continued to pay much higher wates than the once dominant lumber industry, the gap between its wages and those in other regional industries declined between 1911 and 1929. Evidence of an absolute decline in refinery workers' real wages in the late 1920s is contained in a survey compiled by the American Petroleum Institute of wages in the Gulf Coast refineries from 1914 to 1938. This survey reveals a decline in the weekly earnings (in constant dollars) of stillmen and mechanics in the refineries between 1924 and 1929.[42] The success of the American Plan in blocking outside labor organization had apparently weakened the incentive of management to grant increases voluntarily. Of course, wage statistics alone do not include the indirect compensation of refinery workers represented by the wide variety of nonwage benefits introduced in the 1920s. Free medical care, inexpensive housing, and cooperative food stores all increased the workers' total real income. Before the 1920s, these benefits had not been available to any regional workers, and during the decade, they were provided for only those in the large refineries. But wages are an important part of any worker's compensation, and the relative decline in refinery wage rates reflects the basic imbalance of power at the foundation of the American Plan.

Labor historians have usually expanded such general criticisms of the American Plan to include a more specific charge: the benefits freely granted by owners in the prosperity of the 1920s could be freely withdrawn with the coming of the Great Depression. This charge is true. Such, however, probably would have been the case with any arrangement reached in the twenties; the severe dislocations of the 1930s overwhelmed a new, evolving, and as yet unfinished institutional framework for labor-management relations.

More significant than the failure of this cooperative system to survive the depression were the real and lasting improvements it introduced before its demise. A disdain for the primary motive of the owners—the avoidance of independent unions—need not blind the historian to the significance of their actions.[43] The years between 1915 and 1925 brought substantial betterment of working conditions in the refineries on the Gulf Coast. Probably no other decade of the twentieth century witnessed as much improvement in the lot of the refinery worker. For those who entered the refineries before 1915, the comparison of conditions under the American Plan with those of the earlier period served as a powerful rebuttal to arguments that outside labor unions were necessary in order to improve working conditions.

Changes in the regional labor market also militated against the rise of independent unions. Due to the rapid growth of employment in refining after 1915, between 80 and 90 percent of all refinery workers in the Port Arthur area had not experienced the hardships of the formative period. Since the large Houston area refineries were built in the 1920s, all of their employees began work under the American Plan. These workers were able to compare their own lot with that of employees in peripheral industries or with conditions in the cotton, timber, and self-sufficient farming regions in the interior. Indeed, many had migrated from these areas to the coastal region because of the attractiveness of refinery work in comparison to other job opportunities. To such workers, a steady job with relatively good wages and benefits was not an oppression severe enough to justify manning the barricades for the AFL. They had not experienced the grievances of the past. They could not foresee future grievances that would justify the drastic step of attempting to organize a union against the wishes of their benevolent, if paternalistic, employer.

Such grievances came, however, with the Great Depression. Many of the ideas and institutions of the cooperative period were greatly affected by the altered political economy of the 1930s. Economic upheaval shook the existing accommodation between labor and management and reintroduced an increasingly powerful third party, the federal government. In an atmosphere of uncertainty and fear, the three groups sought an arrangement to replace that of the peaceful 1920s.

The business-directed system of the previous period did not vanish overnight. The initial response to the depression, the National Industrial Recovery Act in 1933, produced codes of behavior for the petroleum industry written primarily by the large oil companies. But the 1930s was less a period of business initiative than of business responses to threats. The greatest danger to the traditional economic power of the large oil companies both nationally and within the refining region came from the advance of organized labor. The harshness of the refinery owners' response to this threat was dictated by their perception of its magnitude and by a sense of bitterness rooted in their belief that the workers they once had pampered had now turned against them.[44]

The owners joined the battle against the CIO with an intensity that testified to the depths of their perceived need to retain direct control over their workers. The tactics of the 1920s were not abandoned entirely; wage raises were timed to coincide with organizing drives, and worker participation in company unions was increased to meet demands for independence.[45] But the 1930s were not primarily a period of coaxing, and the oil companies resorted to a type of intimidation that had generally only been threatened in earlier periods. Humble's Baytown refinery at one point resembled a large fortress, complete with an airfield for flying in men and supplies. Local governments, long influenced by the owners of their dominant industries, proved useful allies in harrassing and intimidating union organizers.[46] Advertising campaigns in regional newspapers characterized the CIO as a communist dominated, race-mixing organization whose victory would destroy the area's economy.[47] Within the refineries, widespread discrimination against union sympathizers occurred.[48] Faced with a loss of authority in the refineries and even in the communities that they had helped to build, the refinery owners sought to restore their control with a level of force never before or since utilized in regional labor "relations."

In the past, refinery workers would have been helpless against such a show of power. Before the establishment of the aggressive, CIO-affiliated Oil Workers' International Union in 1936, organizers for outside unions had lost each major economic or political power struggle with the large oil companies. But a unique set of conditions in the late 1930s enabled the CIO to challenge successfully the refinery owners' control. Most important was the changed political environment and the crucial aid given to the CIO by a powerful government agency, the newly created National Labor Relations Board. The NLRB opened plant after plant to elections ultimately won by the CIO.[49] Of nearly equal significance was the financial and organizing aid of the United Mine Workers and the Steelworkers Union.[50] The timing of organization was also propitious. The initial regional success of the Oil Workers' Union came in the period from 1939 to 1941, during the buildup for World War II and the construction of large new catalytic cracking units on the Gulf Coast. The following four years of war-induced cooperation gave the union an opportunity to consolidate its gains before facing a reaction by the oil companies in 1945. With such advantages, the CIO was able to organize many of the industrial workers in the refineries on the upper Texas Gulf Coast (see Chart 6.1). By the end of World War II, the CIO had won its challenge and had emerged as the single most important organization for refinery workers in the coastal region and in the nation.

Several other significant, though somewhat smaller, labor organizations also had emerged by the end of the war. The AFL, which before the late 1930s had proved incapable of establishing a large, permanent membership in the regional refineries, expanded rapidly among numerous crafts from 1939 to 1945. These unions were especially successful in organizing the electricians, machinists, and boilermakers, the largest crafts in the refineries. The AFL groups were often intense rivals of the CIO during the organizing campaigns of the 1930s, but they later

Chart 6.1. OWIU (CIO) National Membership, 1938–1946

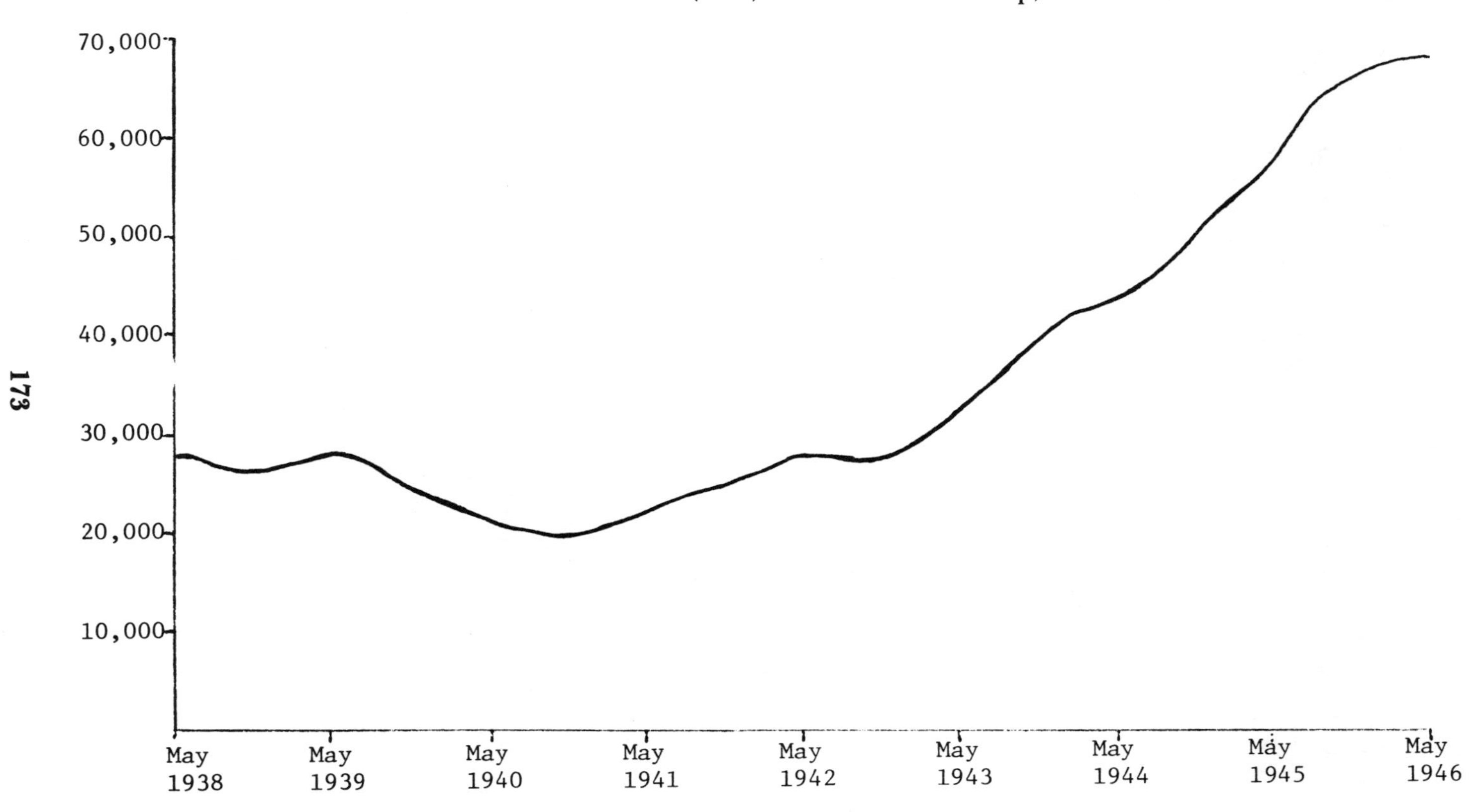

Source: OWIU *Annual Executive Reports.*

cooperated with the Oil Workers' International Union before finally merging with it in 1955.[51]

A second group of CIO rivals, the nonaffiliated "independent" unions, remained separate. The most important nonaffiliated unions in the region were in the plants controlled by Standard of New Jersey, the Humble refinery in Baytown, and the Esso plant in Baton Rouge, which were then two of the largest refineries on the Gulf Coast. Despite several intense organizing drives by the CIO, workers in these plants retained local, cooperative labor organizations that had evolved from the Joint Conferences of the 1920s. In response to changing conditions in the 1930s, Standard allowed these "company" unions to assume an increasing measure of independence, and their members were content to hold on to this familiar arrangement rather than risk the uncertainty of the CIO. This choice was influenced by two related factors. First, Standard's much advertised system of employee relations from the 1920s proved very adaptable, maturing into a labor organization capable of fulfilling many of the needs of its work force. But as in the past, another very different undercurrent also contributed to the federations' survival. Standard had long fought against outside unions, but its reaction to the CIO was especially strong. During World War II, workers at Standard's small Ingleside, Texas, plant near Corpus Christi joined the CIO. After the war, the plant was simply shut down and dismantled. In defense of its action, which of course did away with the first CIO foothold in a Jersey Standard plant, the company cited its pre-war plans to phase out the facility. But to employees in its other, larger plants, the threat was clear. One reason they subsequently avoided the CIO was thus the commonly held assumption that Standard would take harsh measures if necessary to remain free of outside unions.[52]

The increasing independence of the federations, like the growth of the AFL, reflected the indirect influence of the CIO throughout the refining region. That union's well-financed, well-coordinated, and NLRB-aided national organizing drives introduced a new era of independent industrial unionism in the regional labor market, and other labor organizations prospered in the altered environment created by the initial success of the Oil Workers' International Union.

The long-run impact of the "revolutionary" CIO has probably been exaggerated by scholars in search of a "heroic" working class, but in the case at hand, the historical record justifies a heavy emphasis on the CIO's role. The struggles of this period in the refining region were indeed heroic, at least in the sense that they were successful against great odds. Although the Oil Workers' Union did not radically alter the existing social structure, it greatly expanded the power of organized workers in the existing political economy while significantly strengthening the position of the individual refinery worker. A reaction against the hyperbole of previous advocate-historians of the labor movement or a disillusionment with the role of "Big Labor" in the post–World War II economy should not cause the historian to minimize the importance of the early accomplishments of the CIO. The redistribution of power in the refineries between 1935 and 1945 was greater than in any other decade in the coastal region's history. In that period, "outside" unions gave the workers the dignity and independence of enforceable guarantees on matters of job security and working conditions.

A major wave of strikes in 1945–46 proved the permanence of the Oil Workers' Union, which wrested wage leadership from the traditional

Table 6.5. Average Wages and Salaries by Industry, Jefferson County, 1947

Industry	Number Production Workers	Average Wage	Wage Index	Number Salaried Workers	Average Salary	Salary Index
All industries	18,851	$3209	100	4330	$4016	100
Petroleum and coal	13,737	3526	110	3006	4285	107
Chemicals and kindred products	726	2930	91	224	3920	98
Printing	241	2979	93	209	2785	69
Miscellaneous manufactures	17	2765	86	1	1000	25
Primary metals	241	2378	74	28	4768	119
Fabricated metals	2,518	2322	72	220	3741	93
Food and kindred products	743	1863	58	414	3138	78
Lumber and products	684	1806	56	29	4793	119
Stone, clay, and glass products	103	1767	55	10	3600	90
Furniture	82	1707	53	14	2286	57

Source: Census of Manufactures, 1947, vol. 3, p. 591.

Table 6.6. Average Wages and Salaries by Industry, Harris County, 1947

Industry	Number Production Workers	Average Wage	Wage Index	Number Salaried Workers	Average Salary	Salary Index
All industries	46001	$2859	100	12605	$4028	100
Petroleum refining	8326	3987	139	2520	4317	107
Chemicals & allied prod.	3388	2874	101	1112	3790	94
Paper plants	931	3382	118	250	4072	101
Printing	1771	3274	115	1327	3112	77
Machinery (except electrical)[a]	9044	2992	105	2910	4372	109
Primary metals	3614	2834	99	335	5466	136
Transportation equipment	1023	2814	98	218	4128	102
Electrical machinery	431	2740	96	89	6337	157
Fabricated metals	4156	2537	89	721	3702	92
Miscellaneous manufactures	563	2412	84	119	4143	103
Stone, clay, glass products	1407	2227	78	153	3497	87
Food and kindred prod.	5398	2216	78	2238	3641	90
Instruments and related products	355	2177	76	60	2917	72
Textile mill products	559	2077	73	42	5214	129
Furniture and fixtures	965	2066	70	140	5207	129
Lumber and products	2034	1984	69	142	3810	95
Apparel and related prod.	1747	1501	53	193	4295	107

Source: Census of Manufactures, 1947, vol. 3, p. 589.

[a]Includes 7553 production workers in oil field machinery.

industry trend setter, Jersey Standard.[53] This was the climax of a long-run change from direct oil company control of its work force to shared control between the companies and the increasingly independent workers' organizations. In the refining region, the contingencies of World War I had accelerated this change by establishing important precedents concerning workers' rights to negotiate collectively with management. The period from the end of World War I to 1945 was then characterized by two related trends: the growing independence of labor organizations and the increasing influence of national forces on labor relations in the large refineries. The most significant national development from the employers' perspective was the American Plan, which established an accepted model of enlightened treatment of labor against which the employer was judged by the public, as well as the labor force. For the workers, the most important national factors were the rise of the CIO and the pro-labor actions of the early NLRB.

Some of the results of this long-run movement toward independence are apparent in Tables 6.5 and 6.6. By 1947 refinery workers had reopened the gap between their wages and those of most of the remainder of the work force. The independent power of their unions accounted for a portion of this gap. Refinery wages had, however, always been relatively high, and further increases were not the most important advantages of independence. More significant were changes in the realm of job security and job control. The powerlessness of the pre-World War I era and the subservience of the American Plan had been replaced by union enforced guidelines on tenure, promotion, seniority, and numerous other job-related issues. While retaining and even improving their material conditions, the independently organized refinery workers had wrested a large measure of control over their jobs from the owners of the refineries. This was their most significant victory, for it permanently altered the institutional framework within which decisions affecting labor were made.

THE ERA OF RAPID TECHNOLOGICAL CHANGE AFTER 1945: THE QUIET REVOLUTION

The voilent confrontations that accompanied union organization attracted public and scholarly attention, but throughout the century a quieter revolution was in progress. Changing techniques of production gradually but fundamentally altered the worker's position within the plants. The substitution of capital for labor greatly affected the individual worker, the refinery unions, the regional labor market, and the course of regional development. In the post-World War II era, rapid technological change steadily undermined the power of the independent unions, in a sense reducing the significance of their previously hard-won victory.

Other factors, however, tended to solidify the position of the refinery unions. Periodic strikes after 1945 established the power of the unions to conduct orderly, disciplined walkouts capable of shutting down even the largest regional plants. This was made easier by the expanding power of national labor organizations, especially the Oil Workers' International Union (CIO). After the merger of the CIO and AFL in 1955, refinery workers were represented by the Oil, Chemical, and Atomic Workers (OCAW), a labor organization that matched both the international scale and the diversified scope of the largest oil companies. The regional power of this organization was greatly enhanced by the inclusion of most of the workers in the region's growing new petrochemical plants.[54] The nonaffiliated Employees' Federations, which were active primarily in Standard of New Jersey's plants, also sought to strengthen their bargaining positions through unified, national actions against that company.[55] Throughout the region, refinery workers' unions gradually expanded their control over working conditions.

Such gains did not lead to the economic decline that the owners had gloomily predicted during the struggles of the 1930s. In fact, the post-war settlement proved capable of fulfilling management's primary goal, maintaining a stable labor supply, without the control required by earlier labor-management relations. The arrangement had a built-in measure of uncertainty, the threat of strikes during bargaining periods. But these strikes were at least not as random as before; they were predictable, and preparations could therefore be made for them. The owners did not discard the rhetoric of the open shop, but they learned to live and prosper with union labor. Under the post–World War II accommodation, labor also achieved its primary goals, improved conditions and job security. Wages continued to increase both absolutely and relative to those of other regional workers (see Tables 6.7 and 6.8). The final interested party, the government, was able to remain largely uninvolved during the post-war period. In the 1960s, the federal government did impose one significant alteration in the refinery labor force, the legal abolition of the black labor gang and other racially discriminatory employment practices.[56] But the government's post-war role was generally restricted to that of a mediator in the crises caused by strikes, and these

Table 6.7. Average Wages and Salaries by Industry, Jefferson County, 1963

Industry	Number Production Workers	Average Wage	Wage Index Number	Other Workers	Average Salary	Salary Index
All industries	18,760	$6660	100	6133	8491	100
Chemicals and allied	2,398	7359	110	729	8335	98
Petroleum and coal	11,553	7183	108	3106	9462	111
Transportation equipment	1,460	5890	88	203	7220	85
Fabricated metals	940	5639	85	275	7393	87
Machinery	446	5471	82	268	6549	77
Printing	343	4287	64	257	5510	65
Food and kindred prod.	677	3916	59	517	5691	67

Source: Census of Manufactures, 1963, vol. 3, Part 44, pp. 21-22.

Table 6.8. Average Wages and Salaries by Industry, Harris and Galveston Counties, 1963

Industry	Number Production Workers	Average Wage	Wage Index Number	Other Workers	Average Salary	Salary Index
		Galveston				
All industries	6,921	$7061	100	2,583	$ 8782	100
Petroleum refining	1,528	7893	112	478	10425	119
		Harris				
All industries*	62,580	$5663	100	34,937	$8234	100
Chemicals and allied	6,194	6994	124	3,920	9054	110
Primary metals	6,493	6500	115	2,087	8808	107
Machinery, except electrical	10,193	5715	101	4,994	8015	97
Transportation equipment	1,555	5415	96	426	6939	84
Fabricated metals	9,342	5401	95	3,550	7114	86
Printing and publishing	3,143	5359	95	2,212	5984	73
Stone, clay, glass products	3,735	4672	83	1,033	6776	82
Food and kindred products	5,975	4626	82	4,227	6455	78
Lumber and products	933	3759	66	189	6561	80
Apparel and related	1,178	3121	55	229	6166	75

Source: Census of Manufactures, 1963, vol. 3, Part 44, pp. 19-21.

*No listing for petroleum refining.

crises were mild in comparison to those during the two world wars and the Great Depression. Because these arrangements suited the basic needs of all three parties involved in the center economy, they proved more durable than previous systems of labor-management relations.

Throughout the post-World War II period, however, technological change within the refineries was affecting the balance of power underlying this accommodation. Such change had occurred almost continuously since the construction of the original refineries after the Spindletop discovery, but it was most pronounced after the late 1930s.[57] The growing independence of labor unions was one of several strong incentives for the acceleration of the pace of technological change. Table 6.9 shows that labor cost per barrel of oil refined declined steadily in the 1940s and 1950s and remained throughout a relatively small percentage of total refining costs. It was, nonetheless, the most unpredictable and uncontrollable refining cost in the post-World War II period. In the 1950s and 1960s, one response to this situation was automation, which usually accompanied refinery expansion and remodeling. This ongoing, accelerat-

Table 6.9. Average Operating Costs of U.S. Refineries, 1926–1969

Year	Total Labor Costs (cents per barrel)	Total Costs (cents per barrel)	Labor Costs as Percentage of Total Costs
1926	47.7¢	137.0¢	35%
1930	43.3	122.5	36
1935	50.3	127.1	40
1940	45.5	108.3	42
1947	41.0	90.0	46
1952	31.2	67.4	46
1957	17.6	51.0	35
1962	14.1	41.6	34
1967	14.8	44.8	33
1969	15.6	47.8	33

Source: *Petroleum Facts and Figures, 1971,* p. 209.

ing substitution of capital-intensive for labor-intensive techniques of production was perhaps the most important process affecting the post–World War II labor market in the refing region. It helped to reduce both the number of workers and the degree of skills required to operate the refineries.

The impact of technological change was directly felt by the refinery workers and their unions. Whereas growth of capacity previously had resulted in more refinery jobs, expansion after the 1950s often required fewer new workers.[58] In fact, the phasing out of obsolete jobs often more than made up for the increased manpower required for the sustained growth of refining capacity. Such change considerably slowed the growth of refinery workers' unions. Had technological advances not altered the ratio of barrels per day to workers employed that existed in Jefferson County in 1929, almost 41,000 refinery workers would have been required to produce the approximately one million barrels per day which was refined in 1963. Instead, only about 14,700 were needed—the refinery work force had increased by only 21 percent in a thirty-four year period in which refining capacity had grown by 233 percent.[59] As the total number of workers required to operate a refinery declined, the number of production workers, the traditional bulwark of industrial unionism, also decreased relative to both plant capacity and nonproduction workers (see Table 6.10).[60] Both trends eroded the unions' position vis-à-vis that of management.

While slowing the growth of refinery unions in general, technological change also influenced the relative growth rates of the different types of labor institutions in the refining region. In the late 1950s several major strikes resulted after management tried, in a variety of ways, to rearrange and reduce their work forces. Technological change had produced a form of refinery featherbedding, with outmoded craft designations hampering what the companies felt was the most efficient deployment of workers. In 1959 Gulf Oil tried to alter the job structure at its Port Arthur refinery to reflect more closely the existing level of technology; in response, the OCAW struck. During the strike, this CIO union received the support of the nonaffiliated federations of Esso's Baton Rouge refinery and the Humble plant at Baytown. The OCAW succeeded in gaining a compromise that slowed the pace at which the work force was reduced and reassigned to new jobs, and Humble employees, who were being threatened by the same sort of efficiency drive, took note. In August 1959 they voted to join the OCAW, in part because they felt that the OCAW would be more successful than their traditional Employees' Federation in blocking the restructuring of job classifications. The OCAW's victory at Humble was its most significant gain in the post-World War II years, since Humble was the largest plant in the Houston-Port Arthur area, and it had previously been the only major refinery in the region not under OCAW contract. In this particular case, efforts to cushion the impact of technological change on the refinery work force thus led to the consolidation of the OCAW's power at the expense of the nonaffiliated Employees' Federation, a change which altered the institutional framework for labor-management rela-

Table 6.10. Plant Capacity and Production Workers at Gulf's Port Arthur Works, 1950–1966

	Capacity (thousands of barrels per day)	Total Workers	Production Workers	Capacity per Production Workers (barrels per day)
1950	230	5172	3848	60
1951	230	6141	4729	49
1952	245	6076	4697	52
1953	245	5978	4651	53
1954	245	5782	4475	55
1955	245	5684	4343	56
1956	245	5880	4363	56
1957	282	6458	4837	58
1958	270	5940	4473	60
1959	279	5904	4009	69
1960	272	5275	3698	74
1961	269	4035	2744	98
1962	269	3766	2557	105
1963	252	3247	2367	106
1964	258	3221	2319	111
1965	276	3450	2374	116
1966	276	3284	2289	121

Source: Edward J. Williams, "The Impact of Technology on Employment in the Petroleum Refining Industry in Texas, 1947–1966." Ph.D. dissertation, University of Texas—Austin, p. 214.

tions by removing the once powerful nonaffiliated Employees' Federation as an effective option to the OCAW.

On balance, however, technological change greatly weakened the power of the OCAW. In their quest for improved compensation and conditions, the primary bargaining weapon of organized workers was their ability to shut down the large refineries. This tactic had been used to gain basic concessions during World War I and again in the 1930s. During the contract disputes of the 1940s and 1950s, the oil workers' unions usually helped shut down a refinery safely before putting up pickets. Technical and supervisory personnel were allowed to remain inside the plant to safeguard the expensive equipment in the plants, which were potentially hazardous even when not operating.[61] When the refinery involved was the primary producer of its owner or when the OCAW shut down all the refineries of a particular company, that company had a strong incentive to settle the strike and get the refinery back into production as soon as possible. By the early 1960s, however, a significant change was evident. During strikes the increasingly automated plants could be kept in partial operation by technical and supervisory personnel.[62] In the 1960s and 1970s, this nonunion work force grew in number and became more and more proficient at operating the refineries near full capacity in the absence of striking union labor. No longer able to stop production, the OCAW found its bargaining position severely weakened.

In the long run, technological change seemed destined to overturn the post-World War II relationship between labor and management by altering the balance of power on which it was founded. In a labor market being transformed by technology, those implementing this change ultimately control both the jobs and the workers who fill them. Temporary compromises gained by organized labor have not altered a basic trend toward smaller work forces; two years after the OCAW's victory at Humble, management there did away with 722 jobs—11 percent of the total labor force.[63] In the long run, such changes could result in the reestablishment of the center firms in oil as the most important labor institutions in the refining region, for they are the primary controllers of the changing techniques of production in the largest refineries.

NOTES

1. Selig Perlman's characterization of American workers as opportunistic and job conscious seems to be an apt description of these refinery workers. See, Selig Perlman, *A History of Trade Unionism in the United States* (New York, 1922).

2. David Potter, *People of Plenty: Economic Abundance and the American Character* (Chicago, 1954), p. 93, summarizes the process at work: "Throughout our history the development of new geographical areas and new segments of the economy has offered those instances of advancement to which we point when offering illustrative proof that our system of equality enables anyone to succeed." For further discussion of the resulting attitudes, see Irvin Wyllie, *The Self-Made Man in America* (New Brunswick, 1954); Herbert Gutman, "The Reality of the Rags to Riches 'Myth,'" in *Work, Culture, and Society in Industrializing America* (New York, 1977), pp. 211-233. The remainder of the essays in Gutman's book are generally excellent, especially "Work, Culture, and Society in Industrializing America," which

provides a sweeping overview useful in placing events in the refining region in a broader context.

3. Census reports for 1900 through 1930 show that there were very few eastern European immigrants in the region. For a discussion of their impact on the organization of workers in the steel industry, see David Brody, *Steelworkers in America: The Non-Union Era* (Cambridge, 1960).

4. Texas Bureau of Labor Statistics, *Second Biennial Report of the Bureau of Labor Statistics, 1911-1912* (Austin, 1913), p. 53.

5. *The Magpetco* was a monthly plant magazine published in the 1920s at Magnolia's Beaumont refinery. There is a set of *The Magpetco* in the library of the refinery. *The Magpetco,* volume 1, number 1 (April 1921), p. 13.

6. Texas Bureau of Labor Statistics, *First Biennial Report of the Bureau of Labor Statistics, 1909-1910* (Austin, 1910), pp. 133-137.

7. As early as 1904, Texaco faced difficulties in dealing with union bricklayers during the construction of its Port Arthur refinery. See Holmes (Port Arthur) to Cullinan (Beaumont), letter dated March 15, 1905 in *Texas Company Archives,* volume 25, p. 103. For a discussion of some of the unions in Houston, see Robert S. Zeigler, "The Workingman in Houston, Texas, 1890-1914," doctoral dissertation, Texas Tech University, 1972, pp. 141-233.

8. For a history of the AFL in this period, see Philip Taft, *The American Federation of Labor in the Time of Gompers* (New York, 1957), and *The American Federation of Labor From the Death of Gompers to the Merger* (New York, 1959). See also Lloyd Ulman, *The Rise of the National Trade Union* (Cambridge, 1955).

9. An excellent study which places the post-war violence of the steel industry into a broad context is David Brody, *Steelworkers in America.* See also Graham Adams, Jr., *Age of Industrial Violence, 1910-1915* (New York, 1966).

10. For a compilation of early Texas labor laws, see State of Texas, *Laws of Texas Relating to Labor* (Austin, 1909).

11. Port Arthur *Herald,* March 8, 1902, p. 1.

12. In the 1920s, Texaco cited the demise of the line-up system as evidence of improving working conditions. See the *Texaco Star* 13, no. 3 (March 1926), p. 13. For a pathbreaking study of the power of foremen around the turn of the century, see Daniel Nelson, *Managers and Workers* (Madison, Wisconsin, 1975), pp. 34-54.

13. Texas Bureau of Labor Statistics, *Fourth Biennial Report of the Bureau of Labor Statistics, 1915-1916* (Austin, 1917), p. 34.

14. The timber region of nearby East Texas, which was populated largely by subsistence farmers, has been labeled "the land of deep poverty." See Ruth Allen, *East Texas Lumber Workers* (Austin, 1961).

15. Texas Bureau of Labor Statistics, *Fourth Biennial Report of the Bureau of Labor Statistics, 1915-1916* (Austin 1917), pp. 30-64, reports an eight hour per day, six day per week schedule for the regional refineries. Other industries listed generally have nine- and ten-hour days. Paid vacations, while granted sporadically in the early years, did not become common in regional refineries until the early 1920s.

16. Arnold Schlaet (New York) to W.L. Mellon (Pittsburgh), letter dated June 11, 1907, in Drawer A-3, Papers of Joseph Stephen Cullinan, Gulf Coast Historical Association Archives, Houston Metropolitan Research Center, Houston Public Library.

17. For a suggestive discussion of such "dual" labor markets, see Robert Averitt, *The Dual Economy* (New York, 1968), pp. 129-154.

18. Convincing support for this conclusion is found in a study of another center firm: Robert Ozanne, *Wages in Practice and Theory* (Madison, 1968). After a careful study of wage trends at McCormick and International Harvester from 1860–1960, Ozanne concludes that the threat of unionization and then the existence of unions accounts for a substantial variation in rate of increase in wages, a variation which is not accounted for by labor market conditions (pp. 127–129). In the refining region before the organization of the CIO, the threat of independent union activity generally provoked a response by management in the form of an increase in wages or benefits.

19. The largest regional refineries kept pace with the rapid technological change which accelerated with the widespread introduction of thermal cracking. The importance of low labor costs in allowing the refiners to pay high wages in order to block unionization is especially evident in comparison to the situation in the nearby timber industry. See Ruth Allen, *East Texas Lumber Workers,* pp. 185–186.

20. For the average cost of labor from 1926–1969, see "Average Operating Costs of U.S. Refineries, 1926–1969" in American Petroleum Institute, *Petroleum Facts and Figures,* 1971 edition (Washington, 1971), p. 209.

21. George Gibb and Evelyn Knowlton, *The Resurgent Years, 1911–1927* (New York, 1956), p. 143.

22. This is most easily explained by the fact that before its dissolution in 1911, the Standard Oil Trust was by far the largest employer in the industry.

23. For Gulf, see *Oil and Gas Journal,* August 12, 1915, p. 14. For Standard, see *Oil and Gas Journal,* August 19, 1915, p. 14. For Texaco, see *Oil and Gas Journal,* August 17, 1915.

24. Joseph S. Meyers (Commissioner of Conciliation—Houston) to W. B. Wilson (Secretary of Labor—Washington), letter dated June 30, 1917, File number 33-489, "Strike at the Plants of the Gulf Refining Company and the Texas Oil and Refining Company, Port Arthur, Texas," in the Records of the Federal Mediation and Conciliation Service, Record Group 280, National Archives—Federal Records Center, Suitland, Maryland.

25. This was the opinion of the federal mediator involved. See, Ibid.

26. The newspaper clipping from the *Port Arthur-Beaumont Labor Dispatch* can be found in File 33-489.

27. *Beaumont Enterprise,* June 28, 1917, p. 1. See also the summary in the clipping from the *Port Arthur-Beaumont Labor Dispatch* included in file 33-489.

28. Manager of Magnolia's Beaumont Refinery (Beaumont) to Joseph S. Myers (Commissioner of Conciliation), letter dated September 9, 1918, File number 33-1913, "Magnolia Petroleum Company, Beaumont, Texas," Records of the Federal Mediation and Conciliation Service.

29. Myers (Beaumont) to Department of Labor, letter dated September 1, 1918, File number 33-1913, Records of the Federal Mediation and Conciliation Service.

30. L. J. Stone (Beaumont) to Myers (Fort Worth), letter dated November 11, 1918. File number 33-1913, Records of the Federal Mediation and Conciliation Service.

31. Ruth Allen, *East Texas Lumber Workers,* pp. 183–184. One of the best statements of the ideals of the American Plan from the viewpoint of management is a series of articles that appeared in the *Texaco Star* from March 1926 to December 1926. Written by A. A. Nicholson, employment supervisor of the Port Arthur works, the articles argue that "the secret of quality is contentment of the work force," and then show how Texaco is striving to bring about that contentment. Nicholson was one of the new breed of "personnel specialists" at work in the refineries in the 1920s. For a general discussion of the more scientific management in labor relations, see Daniel Nelson, *Managers and Workers.*

32. For one example, see Irving Bernstein, *A History of the American Worker, 1920–1933, The Lean Years* (Boston, 1960), pp. 185–189. Such an emphasis is also evident, especially in periodization, in several histories of oil workers. See Harvey O'Connor, *History of the Oil Workers International Union (C.I.O.)* (Denver, 1950); Herbert Werner, "Labor Organizations in the American Petroleum Industry," in Harold Williamson et al., *The American Petroleum Industry, 1899–1959, The Age of Energy* (Evanston, Indiana, 1963), pp. 827–845; Walter Galenson, *The C.I.O. Challenge to the A.F.L.* (Cambridge, Mass., 1960), pp. 409–426; and Joe Pratt, "The Oil Workers International Union's Organization of the Upper Texas Gulf Coast: 1935–1945," senior thesis on file in the history department of Rice University, Houston, Texas, 1970.

33. The evolving organizational structure of the large corporations is discussed in Alfred D. Chandler, Jr., *The Visible Hand: The Managerial Revolution in American Business* (Cambridge, Mass., 1977). For the oil industry, see John McLean and Robert Haigh, *The Growth of Integrated Oil Companies* (Boston, 1954).

34. For a description of oil field strikes in 1917, see Ruth Allen, *Chapters in the History of Organized Labor in Texas* (Austin, 1941), pp. 221–245.

35. For a brief description of Jersey Standard's early public relations efforts, see George Gibb and Evelyn Knowlton, *The Resurgent Years, 1911–1927,* pp. 249–259.

36. Ibid., pp. 570–595. Henrietta Larson and Kenneth Porter, *History of Humble Oil and Refining Company* (New York, 1959), pp. 209–213. Also, Walter Rundell, Jr., *Early Texas Oil: A Photographic History, 1866–1936* (College Station, Texas, 1977).

37. These benefits varied by company. They are discussed in the *Magpetco,* Magnolia's Beaumont refinery publication; the *Texaco Star,* the *Look Box* (a monthly publication of Texaco's refining department); and in Larson and Porter, *History of Humble Oil,* pp. 209–213. The eight-hour day, paid vacations, health plans, death and injury benefits, and safety programs which were institutied in this period (1915–1925) have all become accepted practices in the industry.

38. *Magpetco* 6, no. 5 (April 1921), p. 27.

39. *Texaco Star* 13, no. 3 (March 1926), p. 13. For a summary of the various employee benefit programs which constituted Texaco's "striving," see Texaco, *Labor-TNEC Submittals,* pp. 3–13 in the Texas Company Archives.

40. Three of the most frequently cited historians share a similar view toward labor in the 1920s. Joseph Rayback, *A History of American Labor* (New York, 1959), labels the 1920s a "decade of decline." Henry Pelling, *American Labor* (Chicago, 1960) depicts labor as "suffering from prosperity." Irving Bernstein, *The Lean Years,* embodies his approach to the 1920s in his title. All write of "labor," but focus on those workers represented by independent, "outside" unions. In the refining region and in other center-related regions, the 1920s were lean years for union organizers but not for those who worked in the refineries.

41. Henrietta Larson and Kenneth Porter, *History of Humble Oil,* p. 209.

42. A.P.I., "Historical Trends of Wage and Hour Conditions in the Petroleum Industry," in *Topics Suggested by the TNEC*—Texaco's TNEC Submittals, p. A-4-8, Texas Company Archives.

43. Of the cooperative institutions which grew in the 1920s, perhaps no other has been so uniformly maligned as the "company union." As scholars attempt to reinterpret the 1920s as something more than a prelude to the Great Depression and Franklin Roosevelt's New Deal, the company-dominated labor organizations merit reexamination as a serious, though aborted, attempt at a cooperative labor-management understanding. See Ellis Hawley, "Herbert Hoover, the Commerce Secretariat, and the Vision of an 'Associative State,' 1921–1928," *Journal of American History* (June 1974), pp. 116–140.

44. The early years of the depression brought conditions which fed a mutual distrust between worker and manager. As a result, once "shared" assumptions gave way to misunderstanding. At Humble's Baytown plant, for example, management introduced a "share the work" program which it saw as beneficial to the welfare of its workers, who, in turn, labeled the program demeaning, unnecessary, and exploitative. See Henrietta Larson and Kenneth Porter, *History of Humble Oil,* p. 359.

45. *National Petroleum News,* April 14, 1937, p. 3, records one example of this tactic. For a more detailed treatment of the organizing drives of this period, see Joe Pratt, "The Oil Workers International Unions' Organization of the Upper Texas Gulf Coast." Attempts to make "company unions" more truly representative were most apparent at Humble's Baytown plant, where the Employees' Federation flourished during OWIU attempts to organize in 1937.

46. Harvey O'Connor's *History of the O.W.I.U. (C.I.O.)* describes the role of the Port Arthur police chief, who had previously been employed at Gulf, in harrassing union organizers. O'Connor is also the source for the description of the Humble refinery. His work is an "official" history of the Oil Workers' International Union which takes an aggressively pro-union view of these events.

47. In Port Arthur, these advertisements were paid for "by people interested in the future welfare of Port Arthur." They argued that the CIO would bring "strikes that cost lives . . . the kind that bring bloodshed . . . the kind that destroy property." See the *Port Arthur News,* July 25, 1938.

48. This discrimination is documented in the hearings before the National Labor Relations Board, which ruled on cases brought by workers alleging unfair treatment.

49. The normal process was as follows: union organizers formally charged a specific refinery with unfair labor practices, usually the maintenance of a company-dominated union. The NLRB then ruled on the charge. In the first six years of its existence (1935–1941), the board was generally pro-CIO, so the ruling usually ordered the company union disbanded and called for supervised elections to determine who should represent the workers involved. After losing a decision before the NLRB, management could appeal to the federal courts. The process was thus time consuming, but ultimately very useful for CIO organizers.

50. The Petroleum Workers' Organizing Committee (formed in 1937) realized that "the Gulf Coast is the key to the oil industry, and until we crack them, we will not be able to say we represent the workers in the oil industry." Thus much of the resources of the CIO-backed PWOC were devoted to Gulf Coast efforts. See Walter Galenson, *The C.I.O. Challenge to the A.F.L.,* p. 415.

51. For a general description of this process in numerous industries, see Walter Galenson, *The C.I.O. Challenge to the A.F.L.*

52. A brief review of trends in post–World War II labor-management relations is found in Henrietta Larson, Evelyn Knowlton, and Charles Popple, *New Horizons, 1927-1950* (New York, 1971), pp. 609-627. For the Ingleside episode, see Henrietta Larson and Kenneth Porter, *History of Humble Oil,* p. 351–389, 600–605.

53. In 1945, Humble granted a 15 percent hourly wage increase and a 22 percent decrease in hours to its Employees' Federation. After a lengthy strike punctuated by President Truman's intervention (the navy took over several of the region's largest refineries), the OWIU gained an 18 percent increase for its members.

54. One significant exception was at the DuPont chemical plants built in Beaumont and Orange after World War II. These large plants had no unions. See Julius Rezler,

"Labor Organization at DuPont: A Study in Independent Local Unionism," *Labor History* 4 (Spring 1963), pp. 178–198.

55. F. Ray Marshall, "Independent Unions in the Gulf Coast Petroleum Refining Industry—The Esso Experience," *Labor Law Journal* 12 (September 1961), pp. 823–840.

56. For a general discussion of this problem, see Anthony H. Pascal (ed.), *Racial Discrimination in Economic Life* (Lexington, Mass., 1972).

57. Three ongoing improvements in refining techniques have had the greatest impact on the refineries' work force: (1) the development of continuous distillation processes to replace batch processes; (2) the construction of large cracking units that increased the yield of the most valuable petroleum products; (3) the introduction and increasing sophistication of automatic, mechanical controls for the refining process.

58. Edward J. Williams, "The Impact of Technology on Employment in the Petroleum Refining Industry in Texas, 1947-1966," unpublished doctoral dissertation, University of Texas—Austin, 1971.

59. These figures are taken from the United States Census of Manufactures for 1929 and 1963.

60. F. Ray Marshall, "Independent Unions in the Gulf Coast Petroleum Refining Industry," p. 829, shows that in the five-year period from 1953 to 1958, the number of production workers in refineries throughout the nation declined by 14.8 percent while production increased by 9.8 percent.

61. See, for example, the daily newspaper accounts of the twenty-five day walkout at the Port Arthur refineries of Texaco and Gulf in May 1952. *Beaumont Enterprise,* May 25, 1952, p. 1.

62. *Port Arthur News,* January 2, 1962, p. 1, reported the conclusion of a seventy-two day strike at the Gulf plant at Port Arthur with the following quotation:

> The big refinery was reported to have been operated at over 50% capacity, with all chemical units except one which was not operated, working at peak production. It was said to have been the first time on so large a scale in the history of the industry that supervisory personnel successfully ran a refinery behind picket lines.

In August of the same year, the OCAW lost a strike at Shell's Deer Park refinery due to its inability to shut down the plant's operations. See Edward Williams, "The Impact of Technology," p. 119.

63. Ibid., p. 129.

Chapter VII

Oil and the Evolution of the Public Sector

The growth of the largest oil firms generated pressures that forced significant alterations in the structure and the activities of local, state, and national political insitutions. As the private companies expanded, they looked to all levels of government for a variety of public services that were essential to their expansion. While using government, however, these firms were also being pursued by it, since the political system was the arena in which the remainder of society registered its complaints about the increasing size and power of business. In the early years, the uncertainties that resulted from antitrust sentiments threatened the operations of the largest oil companies. Later government regulation of labor-management relations and of pollution further challenged traditional corporate prerogatives. In response, the oil companies sought additional control over their political environment. Because they were among the largest and best organized units in a political system that responded to sustained, systematic interest group pressure, these firms were generally successful in securing services from the public sector while restraining its efforts to regulate the oil industry. In the process, they helped transform weak public agencies into strong institutions potentially capable of asserting control over many aspects of regional and national economic development.

AN OVERVIEW OF THE EVOLUTION OF "OIL-RELATED" PUBLIC INSTITUTIONS

The on-going changes in the structure and power of the public sector and in the relationship between oil and state were extremely complex.

The large oil companies exerted a great deal of political power, but they were neither monolithic nor omnipotent in their dealings with a variety of administrative, legislative, and judicial agencies at the local, state, and national levels. There was a great deal of confusion in the different branches and levels of the public sector as it sought to assert new powers in response to the changes brought by the growth of the oil industry, and this confusion was heightened by jurisdictional disputes that inevitably arose as the locus of power moved from the local and state levels to the national government. In the face of such complexity, a somewhat artificial distinction between the changing structure of the public sector and the changing input of different interest groups into that structure is useful in understanding the broad outlines of the evolution of the public institutions that most affected regional development.[1]

One general trend stands out clearly in the evolution of the structure of the public sector. The activities of the large oil companies helped to create new public institutions while transforming the size and functions of many existing ones. The sustained growth of these companies encouraged them to become more active in politics by increasing their need for a variety of services that could be supplied only by the public sector. Their influence then helped to "modernize" those institutions which dealt with oil-related matters by encouraging the rise of centralized bureaucracies staffed with experts. In many instances, the oil industry initially produced both the experts and much of the information used by these public agencies, which often only gradually developed sufficient institutional resources to allow for strong, independent policy initiatives.

In a path-breaking study of the changing structure of American political institutions in the first decades of the twentieth century, Samuel Hays aptly describes the process of modernization then shaping agencies that dealt with conservation, and his analysis provides useful insights into the general nature of the evolution of the modern public sector in America. Unlike many students of politics, Hays looks underneath the anticorporate rhetoric in which much of the debate on conservation policies was couched. He focuses instead on the way large corporations and influential professional groups helped fashion centralized political agencies staffed with experts who sought to make conservation policy based on a scientific approach to utilizing resources in the most efficient manner possible. In describing the resulting shift of the location of decision making from local and state governments to the national level and from grass roots organizations to specialized, bureaucratic agencies, Hays argues that:

> These upward shifts did not arise out of new political theories or the inherent logic of a proper distribution of governmental powers, but rather from the fact that those who fashioned the new patterns of system and functional organizations sought a framework of decision-making consistent in scope and applicability with the scope of affairs they wished to control.[2]

In the political affairs that most directly affected the refining region, "those who fashioned" generally included representatives of the largest oil companies. As the "scope of affairs they wished to control" grew broader and more complex, they encouraged the expansion and the increasing sophistication of those portions of the public sector that provided services essential to their own continued expansion.

Such services were often as important to the companies as the goods and services supplied by the oil-linked industries described earlier in Chapter V. Services provided by public institutions included, but were not limited to, the provision of social overhead capital to build a modern regional transportation system; the exercise of police powers which for a time helped to stymie the rise of independent labor unions; the creation of public agencies capable of governing the pace of the production of oil and of intervening in international affairs to protect foreign sources of crude oil. The political institutions that supplied these and other services displayed a general evolutionary pattern not unlike that of the economic institutions that were closely linked to the petroleum companies. Very strong ties to the oil industry in the early years gradually became somewhat weaker as these public agencies matured and began to serve more diversified political "markets."[3] These agencies were "promotional" in that they encouraged the growth of the oil industry (as well as much of the remainder of the regional economy), and their expansion was an important part of the evolution of the oil-related portions of the public sector.

It was not, however, the only part. In addition to its promotional policies, the government took on greater regulatory powers in a difficult and often unsuccessful attempt to balance the needs of the oil companies with those of the rest of the polity. In the early years, efforts at regulating market structure with antitrust laws far outweighed and even overshadowed attempts at other types of regulation. Amid the turmoil of the 1930s, a second major regulatory function emerged when the government forcefully intervened in labor-management relations. In the 1960s and 1970s hiring discrimination, pollution, and even the use of certain types of energy became the concerns of strong regulatory institutions at the national level. Although the line between "regulatory" functions and "promotional" functions was not always clearly drawn, the basic distinction was that the former made the government a buffer between the oil industry and the remainder of the society while the latter made the government the oil companies' partner in development.

The oil companies had strong incentives to encourage promotional activities of the government while restraining its regulatory efforts, and they sought to accomplish these ends through several means of influence. Perhaps their most important political bargaining chip was the fact that they controlled much of the productive capacity of the refining region and of the state of Texas. Because they were primary agents of economic growth, politicians and the electorate as a whole, were generally hesitant to challenge them on fundamental issues that might slow the pace of growth. At the national level, the large oil companies were one of numerous significant industries that had substantial political influence, but even here they were special, since their product was essential to national security as well as to the operations of many other large industries.

Such broad considerations combined with several more practical sources of political influence to make the large oil companies a very powerful political interest group. Because of their size and wealth, they were among the primary financiers of the public sector. Although exact figures on the total amount of taxes paid by the oil companies to all levels of government are not available, their tax bill was a major source of revenue to local, state, and national governments, and these tax dollars went a long way toward paying for the various services they received from government. Dollars from the oil companies also helped finance another aspect of political life, the frequent election campaigns of those running for public office. Exact figures are far scarcer here than for tax matters, but there is no question that the oil industry provided a significant source of campaign contributions in both state and national politics.[4] Once public oficials were elected, substantial lobbying assured the companies of input into policy formation. In addition, they even provided some of the public officials. For example, Ross Sterling, a founder and president of Humble, served as governor of Texas from 1931 to 1933.[5] Others from the industry filled various elective positions at all levels of government. By maintaining close contact to electoral politics, the oil firms, which were among the largest economic units in the polity, sought to protect and advance their substantial interest in the content of public policy. They established similar, and perhaps even stronger, contacts with government administrative agencies like the Texas Railroad Commission and the U.S. Bureau of Mines. The large oil companies and their primary trade association, the American Petroleum Institute, were important sources for much of the specialized information, expertise, and personnel needed by the growing body of public institutions which sought to regulate aspects of the oil industry, the energy field, and the economy in general.[6]

Such varied sources of input into the political decision making process gave "Big Oil" a potent voice on oil-related issues, but other organized groups also had potential access to decision makers. The political bargaining process that has characterized American democracy in the twentieth century has been labeled "pluralism" by those who celebrate its ability to resolve conflicts among different segments of society and the "broker state" by other, more critical observers. Theodore Lowi, one of the most influential critics of American pluralism, has shown in some detail how well organized groups have been able to exercise a disproportionate power over public agencies and policy decisions which most affect their interests. Lowi's general description of the political process provides a broad context within which to examine political bargaining in oil-related portions of the public sector, where a loosely defined interest group made up of the largest oil companies has often exerted considerable influence on policy.[7]

The victory of "Big Oil" over other less powerful interest groups in the polity is a compelling, if simplistic, version of the process of political decision-making on many issues of crucial importance to the large oil companies. But a historical dimension on the emergence of the broker state adds a missing element of complexity to this view of political policy making. In the early years of oil development in the refining region many with a stake in the pre-oil economy and others with a fear of the growing oil companies sought to use state laws and political agencies to control the expansion of these companies. These groups had some early success in using the state's antitrust and incorporation laws to slow the growth of the largest oil companies, which gradually learned how to defend themselves against political attacks as well as how to use the political system to their own best advantage. After World War I, antitrust sentiment waned and the oil companies, largely unchecked by other political interest groups, enjoyed perhaps their greatest degree of control over oil policy at both the state and national levels. This period of relatively unchallenged political influence proved short-lived, however. In the 1930s and 1940s, organized labor emerged as a political force that opposed the industry on some issues. Then in the 1960s and 1970s, well-organized "public interest" groups also mobilized popular support capable of challenging the industry on issues like pollution control and energy policy. Industry pressure groups remained "more equal than others" because of their traditional access to policy-makers and because of their experience and knowledge of the oil industry, but more than ever before, they were forced to justify their positions and defend them against those of other interest groups. The resulting diversity of inputs into the public sector's decision-making process and the growing maturity and independence of many oil-related political institutions made the formulation of oil policies more democratic, more complex, and more confusing than it had been in the earlier era of largely unchallenged control by the largest oil companies.

The political bargaining process which fashioned oil policy while shaping the evolution of oil-related public insitutions was often accompanied by strong rhetoric that somewhat obscured the nature of the broker state. Presidents from Theodore Roosevelt to Jimmy Carter have called up the potent image of Big Oil in attempts to influence policy, and other politicians at all levels of government have traditionally relied heavily in electioneering on the castigation of the oil companies, one of the biggest of all "Big Businesses." For their part, oil industry spokesmen have further confused matters by opposing much public policy as a threat to the existence of a "free enterprise" economy based on private competition and a limited role for government while, at the same time, seeking to use the government for their own purposes whenever possible.[8] The rhetoric of "the people versus the interests" or of "free enterprise versus the march into socialism" obscures more than it explains when used as a literal description of either the evolution of oil-related public institutions or the changing relationship of these agencies to the large oil companies.

More useful is a historical approach that acknowledges the difficulties facing the public sector as it attempted to adjust to the changed conditions brought by the growth of the oil industry. Many of these difficulties were designed into the American political system by the founding fathers, who constructed a federal system with weak central powers. Others were unique to the late nineteenth and early twentieth centuries, when government was in a very weak position to deal effectively with the changes that accompanied the rise of the modern corporation. But all continued to hamper the rise of an independent, efficient public sector throughout much of the remainder of the current century.

Perhaps the most basic problem that slowed the emergence of a modern public sector was the lack of adequate resources. At the time of the discovery of oil at Spindletop, the state government of Texas was neither large nor sophisticated. In fact, it was singularly ill-equipped to respond effectively to the changes brought by the growing petroleum companies. It lacked the money, personnel, or expertise to match that of the private sector. In 1901 the estimated budgets for the entire governor's office in Texas was only $23,150. The legislative branch was equally limited. It met biennially for about sixty days, during which times its members, who were only part-time public officials, received a salary of $5 a day. Although the state projected total disbursements of about $3.7 million for 1901, most of this money was budgeted for state educational, correctional, and mental institutions, the court system, and Civil War pensions. All other disbursements came to less than $600,000.[5] With such meager financial resources, the state was at a distinct disadvantage in attempting to govern the behavior of such companies as Gulf Oil and Texaco, much less Standard Oil. The national government was in little better position to deal with such companies. Not until the creation of the Bureau of Corporations in 1903 did it have even the capacity to collect information with which to formulate oil policy. Given such weaknesses, government agencies at both the state and national levels often resorted to politically beneficial attacks on Big Oil, which in this era meant John D. Rockefeller's Standard Oil, in place of informed policy making.

Structural differences between the dominant economic institutions and their political counterparts heightened this imbalance of resources. The modern corporate structure facilitated the coordination of worldwide operations in a single organization. In contrast, political authority was fragmented, with various levels of government vying for control over related matters. Corporations were not constrained by the local, state, and national boundaries that marked the jurisdictional limits of political action, and they could take advantage of the confusion of authority inherent in a federal system, concentrating their political resources on the level of power most susceptible to their influence. Since the public sector seldom responded to the growth of the private sector rapidly enough to assert effective controls, this structural difference remained in spite of the type of political centralization described by Samuel Hays. By the time state laws and institutions became capable of handling many of the problems created by oil expansion, the largest companies had become national firms. National political insitutions matured, only to witness the rise of the multinational corporation as the predominant organization in oil. The scale of operations of the major oil companies thus remained larger than that of the many public bodies that grew in response to their expansion.[10]

Differences in the functional organization of the public and private sectors also hampered the efforts of political agencies to deal effectively with the largest oil companies. Vertical integration required the firms to become expert in most oil-related areas. Because political insitutions faced much more general problems, they were not organized in a way that encouraged the accumulation of such specialized knowledge. The attorney general's office, for example, dealt with all antitrust cases and a variety of other matters, not just the antitrust laws as they affected the oil industry or a specific oil company. Those who had access to the best information could hope to guide public policy, and in certain areas, the state's reliance on the large oil companies for specialized information and expertise severely limited its capacity for independent action.

A final significant handicap affecting the governments' relationships with the oil companies was the difference in time horizons. As they expanded, these corporations developed planning capabilities that looked beyond the pressures of day-to-day events. The exigencies of electoral politics, however, limited the planning horizons of the public sector. Those elected to public offices at all levels could be assured of a tenure of only two to four years. Even public bureaucracies were influenced by the resulting constraint on continuity, since elected officials could alter the general thrust of an agency's policy in a variety of ways, including the replacement of directors or changes in funding.

Because of the numerous problems that hampered the public sector in its efforts to define and implement independent, efficient oil policy, the initial lag between its capacities and those of the largest oil companies was only gradually eliminated. Efficient oil-related public insitutions first emerged primarily in areas where the government served a promotional role, whereas the public sector's attempts at regulation were much more halting. This reflected the continuing lack of independence on the part of the public sector, for, until the modern era, it remained quite dependent on the input of the oil industry in determining policy. Independence and efficiency went hand in hand, and neither has been an easy goal for the evolving public sector.

CASE STUDIES IN THE EVOLUTION OF OIL-RELATED PUBLIC INSTITUTIONS

As with the earlier information on important oil-related industries, the following case studies of the evolution of oil-related political institutions are not meant as a comprehensive history of the expansion of those parts of the public sector that dealt with oil policy. Rather, they are intended to supplement and to amplify the overview presented above by examining the histories of some of the imporant public agencies—both regulatory and promotional—that influenced regional development through their impact on the petroleum industry. Because the major oil companies active on the Gulf Coast were national and international concerns, most of these public agencies were outside of the region, at the state and national level. Although the extent of their impact on the region cannot be directly measured, they nonetheless formed an important part of the institutional framework within which regional development occurred. The following case studies of their evolution seek to illustrate the changing relationship between the oil companies and particular public agencies while providing insights into the general process by which public institutions gradually adjusted to the changed conditions brought by the growth of the petroleum industry.

Regulatory Agencies

The emergence of independent agencies capable of defining and implementing efficient regulations has been a most difficult process that has yet to be completed. Throughout the twentieth century, government sought to assert control over aspects of corporate performance that had previously been controlled by the corporation itself. Antitrust laws made the regulation of market structure a government function in the 1890s; the Wagner Act made the regulation of labor-management relations a public function in the 1930s; a spate of laws in the 1960s and 1970s gave the federal government substantial new powers to control discrimination in hiring, pollution, and energy use. In each case, "public" authority was extended into areas that had previously been "private." Each case was accompanied by bitter struggles as the corporations affected sought to resist what was seen as government encroachment into the private sector. For its part, the government's good intentions in asserting regulatory authority were usually not initially matched by the institutional capacity to exercise this authority. The following case studies of the evolution of the structure and policies of several regulatory agencies suggest their regional impact while examining the general process of the growth of the public sector.

The first attempts at regulating the Gulf Coast oil companies were aimed at the market structure of the oil industry, and these early efforts to define and enforce antitrust laws foretold the great difficulties that the public sector would later experience in seeking to regulate other aspects of the companies' performance. Lawmakers at both the state and national levels faced a bewildering array of problems as they sought to interpret and enforce existing antitrust laws amid the altered economic conditions brought by the discovery of oil. These problems were both attitudinal and institutional; that is, there was both an ambivalence toward antitrust which often tempered policy and a lack of sufficient legal and administrative resources to enforce strictly those policies upon which agreement could be reached. Despite such problems, however, antitrust had a significant short-run impact on the growth of the primary petroleum companies that arose in the coastal industry. In the process, these often bumbling efforts to enforce antitrust laws gave many oil executives a very low opinion of both politics and politicians and a keen appreciation of the need to control the threats that regulatory policies could pose for their operations. Such reactions to antitrust in the early years seem to have affected business-government relations long after the threat of antitrust had subsided.

The strict antitrust law in effect in Texas at the time of the Spindletop discovery had been passed in 1889, one year before the U.S. Congress passed the Sherman Antitrust Act, and it prohibited combinations in restraint of trade while blocking one company from owning stock in another. The general incorporation laws of Texas reinforced the antitrust laws by limiting each business chartered in the state to one particular corporate purpose. Strictly interpreted, these laws forbade vertically integrated operations within one oil company. Of course, the actual enforcement of the law was more important than its wording, and enforcement was generally strict in the first decades of the existence of the Texas law.[11]

Actual enforcement was not systematic, however, in part because the state attorney general's office lacked both trustworthy information about offenses and sufficient resources to seek out and prosecute all offenders. The state's rhetorical commitment to correct "the abuse of corporate privileges" far outran its actual capacity to find and attack such abuses. As of 1901, the Texas attorney general's office, which consisted of the attorney general, three assistants, one stenographer, and one clerk, had an annual budget of less than $18,000. By way of contrast, Gulf Oil's original charter authorized a capitalization of $15 million. Such companies quickly outgrew state boundaries and became national and international concerns. They generally had excellent legal staffs and a strong interest in blocking the enforcement of the antitrust laws. As if the companies' advantages in manpower and legal resources over the attorney general's office were not enough, biennial elections gave the public agency a much shorter time horizon than that of the private companies, making continuity of investigation, much less prosecution, most difficult. The state could look for aid to the national government, but it too had only limited experience in dealing with antitrust and meager resources to commit to this problem. With the creation of the Bureau of Corporations in 1903, Theodore Roosevelt moved to improve the federal government's capacity to investigate conditions in industry, but this did not eliminate most of the difficulties faced by the Texas attorney general.[12]

Also undermining efforts at strict enforcement of antitrust laws was a marked ambivalence toward these laws. Populist sentiment had run high in Texas, and a major concern was the power of the trusts, especially that of Standard Oil, one of the oldest, most powerful, and most visible of the industrial combinations. Its large size and growing reputation for secretiveness and ruthlessness combined to make Standard a popular symbol of the evils of concentration. It was thus not surprising that after the discovery of vast new supplies of oil in Texas many citizens—taking a cue from their politicians and their newspapers—became obsessed with protecting the newfound wealth of Texas from Standard Oil. To some extent, however, the political appeal of the antitrust ideology was undermined by the importance of the economic functions performed by corporations such as Standard Oil. Despite repeated warnings that "the Trust" was attempting to capture control of the Spindletop field, and in spite of the wave of protest accompanying Ida Tarbell's articles on "the Octopus" in *McClures,* those directly involved with the development of the area encouraged Standard to take an active role at Spindletop.[13] An association of producers even sent a telegram to Standard inquiring if, given the large sums of money and expertise required for the development of transportation facilities, it would become a common carrier for the entire field. In a similar vein, the *Oil Investors' Journal* welcomed the growth of the Standard-controlled Burt refinery in Beaumont, which "does not cause any jealousy in the public mind; on the contrary, it lends an element of permanency and stability to the situation which is not unpleasant."[14]

The recognition that large-scale, vertically integrated corporations could accelerate the region's economic development did not eliminate the public's fear of concentrated political and economic power; what resulted was an amalgam of these two conflicting strands of thought. This led to a strained ambivalence toward Standard, Gulf Oil, Texaco, and other growing firms. All did not agree that bigness was necessary for rapid growth.[15] Some went halfway by encouraging the growth of large "Texas" firms like Texaco and Gulf as counterweights to the "monopoly power" of Standard Oil. But many, especially those most directly involved in oil, tempered their support of a broad, thorough application of the antitrust laws with an acknowledgment of the efficiency, the desirability, and even the inevitability of large, integrated companies. This ambivalence existed from the beginning of oil-induced growth. Over time, the political demand for a "free enterprise" economy made up of small competitive firms became less strident, and the desire for "a healthy business climate" conducive to corporate as well as regional growth became more widely voiced.

The situation was most tense in the first decade after Spindletop. Then the business climate was far from healthy for Standard Oil and only a bit better for Texaco and Gulf Oil. A series of antitrust attacks on Standard-controlled companies led to dramatic confrontations between the state of Texas and the Octopus. In the Beaumont area, the focus of legal action was the Security Oil Company, a subsidiary of Standard which operated the second largest refinery in the region until 1909. Texas law forbade the entry of Standard into the state, and Security's ties to Standard were an open secret. Nevertheless, Security's refinery, which was Beaumont's largest industrial establishment, was allowed to operate unimpeded until 1906. In 1906 and 1907, despite protests from Beaumont's political and civic leaders, the attorney general of Texas charged the company with violating the state's antitrust laws. In the face of such political pressures, Standard dropped plans for the expansion of its Beaumont refinery and transferred this projected investment to the more hospitable political environment in Louisiana, where its Baton Rouge works later became, for a time, the largest refinery in the world. Found guilty in Texas, Security was ordered to sell its Beaumont properties and leave the state. John Sealy, a Galveston capitalist, purchased the refinery at a court-ordered auction in 1909. The Trust had been removed—at least from public view. In reality, Sealy, with the knowledge and approval of the attorney general, had acted as a surrogate for the Standard of New York interests who previously had owned Security.[16] Despite this arrangement, several years later a new attorney general indignantly "discovered" the remaining Standard connection with the renamed Magnolia Oil Company. Once again the case went to court; a compromise resulted in which Standard interests retained their controlling shares, yet lost the right to vote their stock.[17] Then, in 1917, Magnolia attempted to forestall further legal difficulties by seeking the opinion of the Federal Trade Commission concerning the possibility of its outright purchase by Standard of New York. The FTC, a national regulatory agency created in 1914 in part to give the government an added capacity to investigate industry with an eye to possible antitrust violations, strongly advised against the acquisition.[18] Finally, in the less trust-conscious political atmosphere of the 1920s, the purchase occurred without government protest. In 1925, after twenty-four years of informal associations made necessary by antitrust considerations, the relationship became official. During this interval, antitrust sentiment had not kept Standard out of the state, but the uncertain political environment had slowed the company's expansion and altered its plans for the construction of a giant refinery in the region.

The ongoing pursuit of Standard Oil indirectly involved Texaco, Gulf Oil, and other large oil companies. Local newspapers, state politicians, and the Bureau of Corporations' much publicized investigation of the oil industry in 1906 linked these firms with Standard Oil.[19] The resulting public image of a broad-based conspiracy to deliver the God-given natural wealth of Texas into the hands of the sinister John D. Rockefeller was used and amplified by small producers during battles with the large companies over pipeline policy, over crude oil prices, and over legislation to broaden the state corporation laws. Size was the primary identifying characteristic of companies in collusion with Standard. The largest firms, Texaco and Gulf Oil, thus became major targets of such antitrust agitation.

Those in public office in Texas came to the problem of devising oil antitrust policy armed with inaccurate information, a faulty view of the economics of the oil industry, and an antimonopoly ideology borrowed with little question from an earlier era of railroad rate fights. Ironically, their outmoded attitudes and information yielded results that ultimately encouraged regional growth and the emergence of a less concentrated market structure for the national oil industry. The state of Texas' vigorous, sustained, and often hysterical harassment of Standard persuaded the combine to move very carefully; in 1902, it passed up a chance to purchase the largest regional company, Gulf Oil, because of the political climate in the state of Texas.[20] Had this merger gone through, it would have altered the site of refinery construction and removed the largest company capable of rapid regional expansion. Such a course probably would have slowed the region's growth rate while at the same time strengthening Standard's near monopoly in the national oil industry. Although the state's harassment thus constrained Standard's actions in Texas, its more lax enforcement of antitrust laws against two growing regional concerns, Texaco and Gulf, allowed them to evade existing strictures against vertical integration. In looking the other way as these companies violated the law, Texas enforcement agencies created economic space in which expansive, miniature "oil trusts" could prosper, thereby encouraging the region's rapid development and the related trend toward oligopoly in the oil industry.[21]

Despite these beneficial effects on the regional economy, the individual oil companies recognized the inadequacies of oil policy founded on an ideology that identified bigness with inefficiency and even with evil. Like the politicans they often held in contempt, oil men viewed events through an ideological lens. The dominant attitude within these firms had its roots in the laissez-faire capitalism of an idealized past: the state should not interfere with private enterprise. In fact, the public sector had a long history of such "interference"; but, like the ideology of the antimonopolists, the managers' view arose from a wishful perception of how things should be or how things had been—not how things were.[22]

From such a perspective, the political system was seen as a source of meddling and, above all, economic ignorance. Experienced oil men viewed with contempt politicans who sought to regulate the oil industry without practical knowledge of its operations. Their disdain was magnified by numerous events which, when perceived from an ideology that stressed "free enterprise," strengthened the image of the incompetence of the state. After an exhaustive study that required significant expenditures of manpower and time from the individual oil companies, for example, the Bureau of Corporations concluded in 1906 that Texaco probably was controlled by Standard.[23] Since Texaco officials had made available adequate information to prove the firm's independence, they felt frustrated and betrayed by the bureau, which was the federal agency most directly involved in collecting information for the formulation of a national antitrust policy. Such episodes reinforced their ideological preconceptions and encouraged oil executives to take a disdainful view of politics and politicians: "The oil business is young in Texas. Its operations are to a degree spectacular and attractive to public attention. . . . It is the kind of shining mark which attracts the attention of the average politicians."[24] Similar attitudes caused some oil men to view numerous issues as the creation of opportunistic, incompetent politicans, not as the legitimate reactions of the remainder of the policy to problems arising from rapid, oil-induced growth.

The laws as well as the lawmakers were a source of misunderstanding. Texaco's attorney reported to its president in 1904 that:

> Our [Texas] anti-trust laws as a rule are changed at every session of the legislature and are as complex and intricate as it is possible for such to be and it is wholly impossible to determine in advance with absolute precision whether in the eyes of the State authorities a particular transaction would be violative of this law or not.[25]

In such an atmosphere of political uncertainty, the government became a very important concern for those in control of the growing oil companies.

Especially in this early period, oil executives simply could not ignore antitrust. In the midst of complex economic problems involving new sources of production and a proposed merger with Gulf Oil, the president and the chief counsel of Texaco made the long journey from Beaumont to Austin in 1906 in order to lobby for a broader corporation law.[26] This was not unreasonable at a time when the "most drastic antitrust act enacted by any State" was being passed. The attorney general enforcing this law vowed to "drive every trust and unlawful combination out of Texas," since their "well known purpose" was to "appropriate the territory of Texas for their greed and exploitation."[27] Viewed against the background of Security's treatment at the state level and the dissolution of Standard by state and federal courts, such laws represented a serious challenge to the growth and even the survival of the large oil companies. Antitrust activity in the period before World War I taught an entire generation of oil men to hear, to believe, and even to fear subsequent antitrust rhetoric, as well as other political threats.

The long-run consequences of such fears were predictable. The large oil companies vigorously responded to political uncertainty with a variety of legal and political resources. Texaco's efforts to secure a broader incorporation law are here instructive, for they illustrate the advantages that accompanied great size, good legal resources, and a long time horizon in the American political bargaining process. Although its initial attempt to secure a more favorable law failed in 1905 and 1906, Texaco continued to lobby. Its efforts finally paid off a decade later with the passage of a similar act that had been written by its general counsel and introduced by a Houston congressman. This "Texaco Act" broadened the legal framework for vertical integration in oil by authorizing a refining company to engage in prospecting and production and by permitting pipeline companies to hold stock in other firms. Even the revised law proved less favorable than the laws of incorporation in other states, and in 1926 Texaco moved its corporate charter to the more hospitable legal environment of Delaware.[28] Between 1905 and 1915, the company had become more powerful and more adept at reshaping the structure that affected it in Texas, but its subsequent growth enabled Texaco—and other once regional oil companies—to evade entirely the Texas corporation law by moving beyond its authority. Such lobbying for specific bills took place against the background of more general efforts to shape public and political opinion on oil-related issues. A form of political overkill often resulted; the large oil companies' sustained efforts at containment were often successful enough to allow them to block consideration of some bills and even to initiate favorable legislation. In general, they proved very adept at containing and sometimes controlling those public institutions that represented potential or real threats to their growth.

The oil companies' strong response to political uncertainty seems to have been one of several factors that contributed to the waning of antitrust rhetoric and action. The expansion of these companies beyond the control of the Texas government made effective state antitrust activity more difficult. The dissolution of Standard in 1911 and the public auction of Security pacified public demands for stronger policy. Then, too, the cooperation of oil and state during World War I helped convince many citizens of the advantages of bigness in oil, an industry shown to be essential to national defense.[29]

Underlying these specific events was the basic ambivalence of the antitrust sentiment; a long period of sustained growth and prosperity eased the way for the gradual public acceptance of the large-scale operations of the oil companies. Especially after the 1930s, the periods of instability that had characterized earlier growth were smoothed out, in part at least, by the emergence of public and private secondary organizations capable of a measure of macroeconomic planning. The decline of antitrust sentiment against the oil companies thus resulted from a variety of causes and reflected a general, though gradual, acceptance of the modern corporation by large segments of the American public.[30]

The political sentiments expressed in the early antitrust campaigns did not, however, disappear entirely. In fact, it is difficult to find a period in the twentieth century when one or more of the largest oil companies was not being threatened by antitrust. In the early 1920s and again in the early 1930s, the large oil companies faced attempts by the state of Texas to reduce their size and power. In the late 1930s, the national government briefly pursued a course that seemed pointed toward antitrust actions against the largest companies. Once again, in the late 1960s and 1970s, amid concern over an energy crisis, antitrust sentiment resurfaced in the national political arena in the form of bills calling for divestiture.[31]

Such recurring attempts to regulate or alter the market structure in oil are testament to the continuing political vulnerability of "Big Oil," the modern era's symbolic equivalent to an earlier era's use of the phrase "the Octopus" as a sort of political shorthand for the abuses of corporate power. They are also testament to the essential failure of antitrust in oil, for in the more than ninety years since the passage of the Sherman Antitrust Act, policy makers have yet to devise and enforce stringent antitrust laws that address modern conditions. Indeed, national laws very similar to the Texas incorporation laws of the 1890s are now being proposed as modern approaches to the regulation of the market structure in the oil industry.[32] Despite its inability or unwillingness to enforce strict antitrust laws, the public sector's on-going efforts in this area have taken a great deal of its scarce regulatory resources while engendering strong opposition and suspicion in the oil industry. This constitutes a rather unimpressive legacy in light of the intense passions and exaggerated rhetoric that have been much in evidence from both sides in the disputes over antitrust in oil.

The emergence of other forms of regulation was made more difficult by the fascination with antitrust, which was often presented as a kind of cure-all form of government action that would forestall the need for other types of regulation by returning the economy to an earlier, self-regulating form. Not until the Great Depression dramatically altered the political and economic climate did another regulatory agency have a major impact on the oil-related core of the Gulf Coast economy. In the late 1930s and 1940s, the National Labor Relations Board forcefully implemented a new government power that had been embodied in the Wagner Act of 1935: the power to regulate labor-management relations. By asserting its authority in areas previously controlled almost exclusively by the management of the companies involved, the NLRB played a crucial role in accelerating the on-going evolution of independent refinery workers' unions.

After the creation of the NLRB in 1935, several years elapsed before it exerted a significant influence on the workers in the regional refineries or in any other portion of the work force. Not until 1937 did the Supreme Court remove any doubt about the constitutionality of the Wagner Act. Immediately thereafter, organizers for the Congress of Industrial Organizations (CIO) launched membership drives in several major industries, including oil. NLRB-sponsored representation elections gave these organizing efforts a focus, as union activists could point their efforts toward a coming election with certainity that the election would occur and that it would be fairly conducted by the NLRB. Indeed, from the CIO's point of view, the NLRB was more than fair in the first years of its existence. It was, in fact, a historical oddity in labor-management relations up to that time, a federal agency with real powers that was a friend of labor.[33]

This friendship was revealed in several ways, both in the region and throughout the nation. The NLRB quickly outlawed traditional strike-breaking tactics such as private police forces, blacklisting of union activists, and yellow-dog contracts requiring workers to disavow union membership as a condition of employment. It went much further in its efforts to assure that workers could organize their own unions free from company interference by disbanding workers' organizations that had been founded or were influenced by management. Once such "company unions" were declared illegal, the NLRB would order and supervise elections to determine which labor organizations would represent the workers in each particular plant. At the same time, the board was busy deciding thousands of "unfair labor practices" cases brought by workers against management, and many of these cases involved discrimination against those attempting to organize for the CIO.[34]

NLRB decisions ordering the abolition of company unions in several of the Gulf Coast refineries opened the way there for the CIO, but the Oil Workers International Union (OWIU-CIO) could not take advantage of these openings until the early 1940s. Representation elections held during the economic downturn of 1938 at the three largest regional refineries—Humble's Baytown plant and Gulf Oil and Texaco's Port Arthur plants—resulted in defeats for the OWIU, but after an economic upturn two years later, all of the largest regional refineries except Humble's Baytown works voted for OWIU representation for their noncraft workers. As discussed in the previous chapter, these election results brought a basic change in the distribution of power over wages and working conditions in the refineries. The NLRB did not cause this change, which was part of a long-run trend toward greater independence for workers' organizations underway well before the 1930s. But it did intervene in a way that hastened the process of change. The OWIU faced a long, uphill battle to wrest control over wages and working conditions from the owners of the refineries, and the NLRB evened the odds a bit by defining procedural rules that favored the CIO and then using its powers to assure a "fair fight" under these rules.[35]

The response of the oil companies was similar in tone to much of their earlier opposition to antitrust. The NLRB was seen as a handmaiden of the CIO, which was frequently labeled a communist menace and depicted as a grave challenge to the survival of the free enterprise system. The title of a *Fortune* magazine article in 1938 summarized the prevailing sentiment of the oil companies toward the NLRB: "G.D. Labor Board." From their perspective, this was certainly an apt phrase, at least in the late 1930s, for the NLRB aggressively asserted authority over issues that had previously been the private domain of management. Those who ran the refineries saw the labor board's decisions as threats to corporate autonomy—which they were. Management also feared that the outcome of NLRB intervention would be a strong, independent labor movement that would radically alter the American politic economy. On this, they were only half right.[36]

The OWIU and its successor, the Oil, Chemical, and Atomic Workers, did grow and prosper, and it was most difficult to deal with in the 1950s. But even during this decade of prolonged and often bitter strikes, the NLRB's regulations and its procedural guarantees served as an important check on the return of the unbridled labor "wars" of the 1930s. The board's original mandate under the Wagner Act was "to diminish the causes of labor disputes burdening or obstructing interstate and foreign commerce," and it came to serve this function well. Its most prominent historical role had been its encouragement of the CIO in the late 1930s, but after World War II it became a much more impartial arbiter whose primary function was to set and enforce ground rules for labor-management relations. Its regulatory role as a mediator between those who managed the refineries and those who worked in them thus ultimately promoted the interests of the oil companies by making relations with labor more predictable while generally containing labor disputes in a well-defined legal arena with rigidly drawn and enforced rules. Despite their dire predictions to the contrary in the 1930s, the oil companies, like other major industries, learned to coexist quite well with a strong, independent labor movement in the modern era, and the NLRB helped bring about this significant shift in labor-management relations.

A final wave of regulatory reforms in the 1960s and 1970s brought fresh challenges to corporate control in several new areas. The Equal Employment Opportunity Commission, a federal regulatory agency created to enforce aspects of the Civil Rights Act of 1964, had a dramatic impact on the conditions of labor of minority workers in the refineries. By ending the traditional practice of all black labor gangs, it opened opportunities for advancement for workers formerly consigned to the worst jobs in the refineries.[37] In the 1960s and 1970s, state and national regulatory agencies also vigorously asserted new regulatory powers over pollution control practices in the refineries. Such environmental regulation was made much more difficult by the energy crisis of the 1970s. The desire to conserve oil and natural gas brought new controls on the price of the crude oil refined and on the type of fuel used in the refining process, and these regulations of energy supply often conflicted with earlier regulations aimed at controlling pollution. The emotions engendered by both these new sets of regulations recall those that accompanied both the labor disputes of the 1930s and the antitrust campaigns of the turn of the century.[38] Again, either the end of the free enterprise system was at hand or the greed of Big Oil was subverting the democratic process, depending on whether a spokesman for the oil companies or the regulatory agencies described events. Such was to be expected when the public sector attempted to assert regulatory powers in new areas. That corporate resistance often sounded very similar to much earlier outcries of oil executives against unwarranted government "intervention" should not be surprising, since government initiatives in recent decades were as novel and as threatening from a corporate perspective as the earlier efforts by government to regulate market structure with antitrust laws. The fact that the public sector frequently exaggerated claims of past corporate abuses and promised quick regulatory solutions was also not surprising. Like those who sought to enforce the antitrust laws of Texas in the first decade after Spindletop, the new government agencies that sought to regulate energy and environmental problems had to overcome corporate resistance while attempting to fashion both laws and institutions that would prove effective in the long run—all within constraints imposed by a democratic political system that demanded quick solutions to complex problems.

Chapter VIII explores in much greater detail the historical emergence of environmental regulations as they affected the Gulf Coast refining region, and it will return to many of the themes introduced in this section on regulatory agencies in general. It will develop more fully an overriding conclusion about regulatory activities which this section has only been able to suggest: many attitudinal and institutional problems undermined government's attempt to assert regulatory authority over the oil-related industrial complex on the Gulf Coast without inhibiting its performance as the engine of economic growth.

Promotional Agencies and Functions

Public insitutions encouraged the acceleration of this engine of growth in numerous ways, and the expansion of such promotional activities by the government was generally much easier to accomplish than was the expansion of its regulatory functions. This section will examine three important promotional functions of government that had significant effects on the growth of refining on the Gulf Coast: transportation improvements completed largely by the U.S. Army Corps of Engineers, the stabilization of crude supply accomplished primarily through the efforts of the Texas Railroad Commission, and diplomatic services provided by the U.S. State Department. Although these and other promotional services were of special importance to the petroleum industry, they were often also available to others in the region—as with transportation improvements—or to other center firms throughout the nation—as with the efforts of the State Department. The large oil companies were thus one powerful interest group among numerous others attempting to obtain promotional services from the government, but, at least in Texas, they exerted a strong voice in determining the extent and the distribution of such services from several important public institutions.

The public sector's commitment to improving the region's waterways reflected in large part the general needs of the oil industry and the lobbying of particular oil men. One of the first obstacles that faced the companies spawned by the Spindletop field was the lack of port of entry status for Port Arthur. Galveston, nearly seventy miles down the coast from the fledgling oil refining and shipping center at Port Arthur, housed the nearest customs officials. In the absence of fast, reliable transportation, a trip to the island city of Galveston for the processing of a ship's papers often meant a lengthy and costly delay. Officials for the oil companies pushed hard in the national Congress for the establishment of a port of entry at Port Arthur. Although this seemingly minor change made economic sense in light of the fact that one of the largest oil fields then extant was located near Port Arthur, it ran into heavy and effective counterlobbying from politically well-connected individuals with heavy financial interests in Galveston and in Sabine Pass, a small Gulf port below Port Arthur which stood to lose from its rival's development. Some insight into what the lobbying process looked like from the point of view of those involved in the oil industry is gleaned from the reminiscences of George Craig, a Port Arthur banker with financial ties to John Gates. Craig recalled that Gates had given a $20,000 bribe to buy a Kentucky bluegrass farm for one of Texas' congressmen in a futile attempt to move the port of entry bill through Congress. According to Craig, this bribe had not yielded the expected results because the Secretary of the Treasury had been bribed by the other side in the dispute. Craig's recollections, like most information about such "lobbying," cannot be substantiated, but whether his memory was exact or not, his concluding sentiments aptly reflected those of others in the oil industry who sought favors from the government in this period: "It was once stated that people thought we buy our way through Congress, whereas the real truth is that Congress makes us buy our way."[39] The political system controlled expenditures for transportation improvements and for other social overhead capital which could be crucial to the rapid growth of the oil industry. When such control resulted in a developmental bottleneck, the companies affected sought to remove the political obstructions by gaining political favors by whatever means they deemed necessary.

Simply attaining port of entry status for the new ports that emerged on the Gulf Coast was not the end of the lobbying efforts of the oil companies active in the region. As tanker shipments from the region increased and as the size of tankers grew, substantial improvement of the rivers and harbors along the Gulf Coast was required. Tankers needed deeper water than did most other ships, plus extensive turning basins and special dock equipment. Since their survival depended on access to national markets, the large refiners pushed aggressively for such facilities, and they generally proved successful. Texaco's president, Joseph Cullinan, and one of its major stockholders, John Gates, both lobbied long and hard for federal improvements of Port Arthur's port. Cullinan was later joined by other oil men in lobbying for the completion of the ship channel that made Houston a salt water port. As tankers grew larger and water traffic increased, existing ports and channels were periodically deepened. The needs of the developing southwestern petroleum industry seem clearly to have accelerated an ongoing process of river and harbor improvement.[40] The lobbying efforts of oil men and the needs of the oil industry also hastened the completion of an intracoastal canal which ultimately reached the length of the Gulf Coast from southern Texas to Florida.[41]

The "supply firm" that provided improvements of waterways was usually the U.S. Army Corps of Engineers. In spite of the fact that political lobbying, not dollars, was often required to acquire its services, this public agency performed economic functions no less necessary than those performed by the private corporations that supplied the oil tankers that used the region's improved waterways. The Galveston office of the Corps deepened existing shipping lanes, widened rivers, built jetties to protect the region's ports, helped create an inland, deep-water port at Houston, built the intracoastal canal along the Texas Gulf Coast, and protected life and property during the frequent hurricanes that threatened the region.[42] Some of these activities directly and self-consciously promoted the growth of the regional oil industry, but all were also available to other industries and individuals in the region. The promotional support of this particular public agency greatly facilitated the rapid, sustained expansion of the region's oil-related industrial core by enabling these industries to utilize the most modern means of water transportation available.

The region's rivers supplied more than an avenue for shipping goods;

they also furnished most of the vast quantities of fresh water needed for cooling purposes in the large refineries and petrochemical plants, and public sector agencies insured adequate supplies. In 1917, the assistant manager of Texaco's Port Arthur works placed such demands in the context of the region's overall water needs: "The total water used in operation last year was 6,979,000,000 gallons, enough to supply the city of Houston a year and a half."[43] At the time this refinery represented only about one-quarter of the region's total capacity and, presumably, a similar share of the water used in regional refineries. Private reservoirs on land owned by the oil companies stored cooling water drawn from surrounding rivers, and in this early period and, to a lesser degree, in later years, these reservoirs were among the largest regional sources of water for industrial use. As the sustained growth of regional refining placed an increasing burden on available water supplies in the modern era, special public agencies such as the Lower Neches River Valley Authority were created to insure a "guaranteed supply of water to municipalities and oil and chemical plants," as well as to other regional industries. By the 1970s, however, Jefferson County had to import from surrounding counties over half of its daily supply of ground water. As in the Houston area, the withdrawal of large amounts of water had caused potentially harmful land subsidences.[44] Such problems grew out of the cumulative resource use by a significant portion of the nation's refining and petrochemical capacity and by those who depended on these plants for a livelihood. For most of the century, adequate resources for the region's primary industries had been available, but prolonged and extensive resource use made planning and conservation necessary to assure adequate supplies in the future, and overall planning for all industry required the involvement of the public sector.

Unlike transportation improvements or the supply of fresh water, some government promotional services for the oil companies were not widely available to other industries and individuals. Perhaps the most important such service was that provided primarily by the Texas Railroad Commission, which became a crucial part of the modern political economy of oil. This commission's evolution provides an interesting study in institutional transformation, as an existing agency took on substantial new powers in response to the growth of the petroleum industry. Founded in 1891, its initial mandate had been the regulation of the state's largest pre-oil industry, the railroads. Because it regulated railroad rates, its decisions on the proper rate structure for the shipment of petroleum products vitally affected both the railroads and the young Texas oil companies in the decade after the Spindletop discovery. On several occasions, the Railroad Commission was the scene of vehement disputes between the two industries, and it generally seems to have satisfied the oil companies by holding rail rates on oil down in the early years when the oil companies still depended on the railroads for a substantial portion of their overland shipping. The first president of Texaco voiced his approval of the commission's performance in a letter to a fellow officer of that company in 1904:

> Railroad Commissioner Mayfield joined our party (which included the Governor of Texas), and I had a very pleasant interview with him also. He inquired particularly how the oil rates applied, and whether I considered that the present rates were fair to both the oil interests and the railroads. I told him I thought they were, and that they were reasonable, and if we found them burdensome, we would be glad to let him hear from us at any time, which he stated he would be pleased to do, and to see that careful consideration was given any complaints we might have to offer.[45]

Even in these early years, the Railroad Commission thus acknowledged the economic importance of the oil industry by according it the courtesy of close personal attention.

The ties between the growing oil industry and this particular public agency became even closer after World War I. In 1919, in the absence of a more appropriate state agency to regulate oil production and to issue rules for the conservation of oil, the Railroad Commission took on these new functions. Throughout the 1920s, its new powers were of little consequence, but with the coming of an oil glut in the late 1920s and early 1930s, the commission gained the authority to regulate the amount of oil that could be produced in Texas. The discovery of the huge East Texas field in 1930 threatened to undermine the stability of the petroleum industry by flooding an already demoralized market for oil with vast new supplies of inexpensive crude, and many oil men turned to the government to correct the situation. Amid the chaotic conditions of the early 1930s, the Railroad Commission took charge, and its ability to establish and enforce allowable levels of production in the largest oil producing state in the nation gave it considerable influence in the national and international political economy of oil. A series of "interstate oil compacts" throughout the Southwest pegged the allowable production of surrounding states to that established by the Texas Railroad Commission, and a "Hot Oil" Act passed by the national Congress in 1935 further reinforced the power of the Texas agency by outlawing the interstate shipment of oil originally produced in violation of state prorationing decisions. Taken as a whole, the changes in the public sector in response to crisis conditions in the oil industry in the 1930s made the

Texas Railroad Commission one of the most important public institutions involved in the oil industry from the 1930s through the 1960s. Its stated function was "regulatory": the conservation of petroleum through limits on wasteful production. But its primary practical function was "promotional": the stabilization of the price of crude oil. It served this function well until a very different type of "public" institution, the Organization of Petroleum Exporting Countries (OPEC), asserted its own authority over crude prices in the 1970s.[46]

The operation of the Railroad Commission illustrates how the complexity of the functions performed by public institutions increased along with their scale. In 1905 the United States Bureau of Corporations had attempted to impose federal authority on the industry and had been forced to interview officials from each oil company in order to secure information about the oil industry as a whole; at that time, Standard took care of its own planning, and, in effect, market forces took care of the rest of the industry. By the 1930s, however, the Texas Railroad Commission was becoming a centralized planning agency for the state's industry and, at least in regard to production control, for the industry of the entire nation. In addition to the cooperation of the large companies, it had access to the information gathered by several other centralized, bureaucratic bodies, including the Bureau of Mines and the American Petroleum Institute. Between 1905 and the 1930s, an industry characterized by rapid geographical and technological expansion had developed the combination of private and public institutions it needed to provide detailed and systematic planning, and despite its unlikely name, the Texas Railroad Commission had a central role in performing this function.

Much of this geographical expansion took the once "regional" oil companies across national boundaries, and in the disputes that inevitably occurred with other national governments, the diplomatic services of the United States Department of State became very important to the oil companies. These services were promotional in that they generally backed the claims of U.S. oil companies with the diplomatic resources of the U.S. government. The State Department's activities did not, of course, take place inside the Gulf Coast refining region, but they did directly affect the growth of the coastal refineries by helping to secure foreign crude supplies for use in them. As foreign sources of crude became more and more important to American oil companies in the modern era, diplomatic efforts by the State Department became closely intertwined with the oil companies' unending attempts to retain control of supplies of oil throughout the world. The economic fate of much of the oil-related industrial complex on the Gulf Coast was tied through the vertically integrated operations of the large oil companies to the success or failure of such foreign policy initiatives.

This connection was most evident in the first encounter between the Gulf Coast oil companies and a foreign government, the revolutionary government of Mexico after 1910, and a detailed examination of the interaction of the oil companies and the State Department in this dispute sheds light on their initial relationship. In their early quest for new sources of crude oil, Texaco and Gulf Oil crossed the Texas border into Mexico, where they joined other large American and British companies, including Standard of New Jersey, Sinclair, and the Huasteca Petroleum Company. Much of the crude produced in Mexico was refined in the Texas and Louisiana Gulf Coast refineries of these companies, and several large new refineries were built near New Orleans specifically to handle Mexican crude. In the period from about 1915 to about 1925, the availability of an expanding supply of Mexican crude was thus an important determinant of the pace of growth of the Gulf Coast refining complex. When disputes with the Mexican government threatened this important new source of crude, Texaco and Gulf Oil became involved for the first time with the U.S. Department of State, which was not yet very experienced in dealing with either the revolutionary nationalism of the Mexican government or the diplomatic needs of the modern petroleum industry.[47]

The crucial issue in the dispute between the oil companies and the Mexican government was the control of subsoil rights. During the World War I, the revolutionary government of Venustiano Carranza had enacted a new constitution. Article 27 restored control of subsoil mineral rights to the government, a status that was customary in Hispanic law but had been altered by the regime of Porfirio Diaz (1876–1910), who had granted these rights to American and British oil companies in an effort to attract foreign capital. With the return of peace, these companies sought to prevent the implementation of Article 27, which they considered confiscatory and retroactive. A decade of dispute followed with the fate of the valuable Mexican oil fields in the balance. Given the unwillingness of either the Mexican government or the oil companies to surrender on the basic issue of ownership of the subsoil rights, the crucial variable during much of the debate was the extent to which the American and British governments would push the claims of their national oil companies, and the American companies thus went to great lengths to gain the support of the State Department.

Shared goals, common sources of information, and a close working relationship made the State Department the willing ally of the oil companies. Accepting the interpretation of events presented to it by the representatives of the companies, the department did everything short of armed intervention to force Mexico to yield. In the early 1920s the fear of an impending oil shortage and of British control of international sources of crude oil led the State Department to sponsor the efforts of American oil companies to expand proven sources of crude throughout the world. An aggressive oil policy helped satisfy the State Department's requirements for national defense while facilitating corporate goals of expansion. Removal of the threat to expansion in Mexico, the leading supplier of American crude imports in the early 1920s, was thus an area of special concern for both.

In dealing with the Mexican situation, the State Department's use of information provided by the oil companies strengthened this identity of interests while suggesting a crucial weakness of the State Department in these diplomatic negotiations. Much of State's knowledge of events came directly from the oil companies, reaching Washington by several routes. The oil men maintained excellent relations with the consulate at Tampico, and the consul usually funneled their view of events up to Mexico City and then to Washington. The path taken by a confidential memo prepared by Mexican Gulf Oil Company demonstrates this process. In response to a request by the Tampico consul, Gulf Oil, which had "special facilities for obtaining data of this nature," submitted a "List of Radical Agitators in Mexico," including 338 names of men whose radicalism consisted of membership in local oil workers' unions. The consul simply changed "Bolshiviki" to "Bolshevik," then sent copies of the Gulf Oil report to the State Department, the American embassy, and the consulate general. Listing the source as "a reliable local American concern," the consul suggested that the list would be helpful for visa work, "but may also be useful for other purposes."[48] As association of the American companies in Mexico also provided a steady supply of information such as English translations and commentary on Mexical laws, newspaper clippings, and bulletins on the progress of court cases. Jersey Standard's lawyer held conferences and kept in contact with both the secretary of state and the chief of the Division of Mexican Affairs. Though the oil companies' information was usually reliable, it could also be slanted and at times even distorted, as when Standard prepared a memo for the Division of Mexican Affairs which stated that "probably from time immemorial" hydrocarbon subsoil rights in Mexico had been the property of the surface owner.[48]

The extent to which the State Department followed the lead of the large oil companies best revealed during a dispute over export taxes in 1921, when the secretary of state arranged for a committee of oil men, headed by Standard's president, Walter C. Teagle, to negotiate directly with representatives of the Mexican government.[50] But even though oil and State usually agreed on both the problem at hand the the solution desired, they could not force their views onto the Mexican government. At various times, the United States resorted to nonrecognition of the Mexican government, an arms embargo during a period of civil war, the sending of battleships to Tampico, and a formal diplomatic protest against Mexican legislation which had not yet been enacted into law.[51] This pressure at times caused Mexico to exercise a tactical retreat but never to renounce the intent of Article 27.

This joint offensive by the oil companies and the State Department ultimately failed due to the continued opposition of the Mexican government, but the process at work, not the particular result, is of most concern for the study of the evolution of political institutions. In these early efforts to deal with the conflict between American corporations active abroad and the revolutionary nationalism of a newly assertive "host" government, the State Department was essentially a follower of the petroleum companies, not an independent policy-making institution. Its central weakness, the lack of reliable, independent sources of information, was much in evidence in the excessive reliance of the foreign consul at Tampico on the American oil companies for information about their disputes with the Mexican government and even for interpretation of these events. Because Tampico was located in the center of the oil fields in which these disputes occurred, its consul supplied much of the information that was filtered up through Mexico City to Washington to become a primary input into the foreign policy decision-making process. The State Department was, however, also influenced by the constraints imposed by political opinion, not just by the needs of the oil companies, and in 1927 a Senate resolution passed by a unanimous vote demanded the resolution of the conflicts between American concerns and the Mexican government through continued negotiations, not through military intervention as demanded by some oil men.[52] Thus despite its weaknesses as a policy-making institution and its agreement with the oil companies on the basic ends desired, the State Department faced political constraints on its choice of means which were not faced by the private concerns.

Improvement in the institutional capacity of the State Department to make well-informed, independent decisions gradually altered its status as a follower of business. In response to the growing demands placed on it by the expansion of American industry and citizenry abroad, the State

Department undertook a series of reforms between 1909 and 1922 to rationalize the department in order to make it more efficient as a "diplomatic machine."[53] The subsequent upgrading of the quality of personnel through the introduction of "professional" selection, training, and promotion procedures gradually altered the department's relationship to the large petroleum companies. Before the reforms, foreign consuls, who were crucial sources of first-hand information, had been part-time employees of the department who earned the remainder of their salaries as agents of companies such as Standard Oil.[54] The elevation of the consuls to full-time department employees made their well-being and advancement dependent on performance satisfactory to the State Department, not to private concerns. As it developed into a modern bureaucratic institution, it became increasingly able to define policy with the needs of particular industries as one of numerous influences, not as the only significant influence. In short, in becoming more efficient, it became more independent. In the late 1920s, the new ambassador to Mexico, Dwight Morrow, voiced this change in emphasis while also pointing out the earlier weakness of the State Department:

> The last six months have been quite a disclosure to me of the extent to which responsible oil companies seem to believe it is the duty of the State Department to run their business in foreign lands.[55]

Such a belief should not have been surprising, since the companies had been well trained to expect such treatment by State's earlier behavior. Forging a new, more equal relationship with the large oil companies subsequently proved to be very difficult, however, as their expansion took them from Mexico to other parts of the world—particularly the Middle East—in which the State Department had little or no previous experience. The crucial importance of oil for national security also tempered the State Department's willingness to separate the national interest from that of the large international petroleum companies. Although the State Department gradually ceased to be a follower in defining the "foreign policy of oil," it became instead more a partner than a leader in its relationship with the oil companies.[56]

As the activities of the companies that owned the large coastal refineries became increasingly national and international, the public institutions that most affected them were national agencies. The growth of the large oil companies that owned the refineries created needs for promotional and regulatory activities which the public sector came to address at the national level. The impact of political institutions like the

State Department on the oil industry was then transmitted back to the refining region through the vertically integrated structures of the oil companies. Large, secure sources of crude oil were essential to the sustained expansion of the modern oil industry of which the giant coastal refineries were important parts, and much of the interaction between the oil companies and the government centered around this crucial concern. The Texas Railroad Commission and the State Department thus played especially important promotional roles by helping to guarantee stable supplies of crude.

In two intervals of the twentieth century, World Wars I and II, business and government became even more heavily involved in cooperative approaches to economic planning. These two brief periods had a significant impact on a region whose primary export was an essential product for modern warfare. Because of its relative security from German invasion and submarine warfare, the Gulf Coast refining complex expanded rapidly under the guidance of the Fuel Administration during World War I. Government-business cooperation during World War II lasted much longer and produced far broader economic changes in the Houston-Port Arthur area. Under the impetus of war, joint investments by the federal government and the oil companies significantly expanded the region's industrial capacity while contributing to an important alteration in its industrial mix. Massive aviation gasoline facilities and advanced cracking units added to the refining capacity of the area. Also built during the war were modern petrochemical concerns and entire new synthetic rubber plants. These units utilized the most modern available technology and represented a great leap forward in the capacity and the sophistication of the region's refining and petrochemical production. After the war, the government sold its interest in these plants to existing private companies at prices favorable to the purchasers.

The fact that the government did not retain a portion of these facilities after the war, despite the strategic importance of aviation fuel, synthetic rubber, and petrochemicals, revealed a crucial limit to the state's role in the mixed political economy. During a period of emergency, government investment in producing plants was tolerated temporarily, but the government rapidly withdrew from such involvement as soon as the emergency ended.

This limit had been delineated clearly in debates over a proposal to establish a state oil company in 1914. Secretary of the Navy Josephus Daniels and Under Secretary Franklin D. Roosevelt backed a bill introduced in Congress that would have enabled the navy to build and operate a 20,000 barrel per day refinery and related production and transportation equipment. If passed, this bill would have resulted in the construction of a major new plant in the refining region; at the time, only Gulf's Port Arthur refinery was significantly larger than the navy's proposed facility. The scheme attracted the comment of "over a dozen of the largest operators in the South Texas and Louisiana fields"; all were "emphatic" in their opposition. They predicted that the result would be an "enormous loss to the Government" and "the creation of some fat politicans." Such cynicism from oil men did not dampen the enthusiasm of the real estate agencies and chambers of commerce which sought to attract the new plant to particular locations within the region. But industry opposition and the onset of World War I spelled the demise of the state company, even though extensive investigations as to cost and location were completed.[57]

The furor over Secretary Daniels' "radical plan" provides a useful guage of the sentiment for and against such direct government participation in economic life. Daniels' justification for public ownership—the navy needed a dependable source of fuel at a reasonable price—implied that in this case refined oil products were, in a limited sense, public goods, since they were essential for national defense. His opponents saw the proposal in a different light; they viewed it as an unwarranted and inefficient intrusion of the state into the private sector. The state could deepen harbors which served the refineries; it could build highways which encouraged the use of their principal product; it could mediate labor disputes within the plants; it could even control the quantity and indirectly the price of crude available for their operation. The government could not, however, own and operate a refinery to supply the navy's fuel.

The implications of such a constraint were far-reaching. National defense needs would continue to be entrusted to private concerns, thereby increasing the claims of these firms for government promotional services necessary for expansion, especially into foreign fields. The private corporations would continue to control most of the information, technical expertise, and experience required to understand the workings of the oil industry. Without its own sources of such knowledge, the state would remain handicapped in its efforts to regulate intelligently or independently the growth of the large oil companies, the development of the oil industry, or the nature of the nation's energy mix. Despite oil's central role in the modern economy, the United States, unlike most other industrial nations, would have no single regulatory agency capable of independently overseeing its operations.

Such constraints on the public sector help explain why private corporations retained such a large measure of control over many decisions that vitally affected the society as a whole. Underlying the oil companies' political power was their past record of economic success; they had compiled an excellent record of sustained growth which benefited most segments of the polity. But the primary goal of these companies was not the betterment of society; their organizational imperatives and incentives dictated choices based on their own long-run well-being. When these choices had consequences detrimental to the remainder of the society, those affected had access, at least in theory, to public remedies; but the public sector was often unable to deliver such solutions. The magnitude of the problems resulting from rapid growth sustained over a long period of time in part accounted for this failure. Also important in evaluating the public sector's performance was its inexperience in dealing with oil-related problems. Many of its duties involved new functions for which it had little expertise and only limited resources. In addition, public agencies generally responded to problems only after private solutions had failed. Indeed, the private corporations' ability to impose their own solutions through public institutions often eliminated the opportunity for broader political responses.

The weaknesses of the public sector had far-reaching implications for the growth of the oil industry and of the refining region. At stake was the ability of those within the region to control the process of growth and to respond to its consequences. The political resolution achieved was of central importance in determining the course of growth, since political forces came to shape an ever wider range of economic events. The significance of such forces can perhaps be seen most clearly when viewing specific problems in depth, thus the following chapter, which deals with the environmental problems caused by the growth of refining in the coastal region, provides a case study of the evolution of private and public planning bodies and of the changing relationship between the two. The specific institutional responses to the problems caused by refinery pollution should therefore be seen in the context of the broad political changes wrought by the growth of refining on the Gulf Coast.

NOTES

1. For a general review of the various ways scholars have viewed regulation, see Thomas McCraw, "Regulation in America," *Business History Review* (Summer 1975), pp. 159–183. Also, Richard Posner, "Theories of Economic Regulation," *Bell Journal* (Autumn 1974), pp. 335–358.

2. Samuel Hays, *Conservation and the Gospel of Efficiency* (New York, 1969), preface to the Atheneum Edition (n.p.).

3. For a provocative discussion of these ideas, see Charles Lindblom, *Politics and Markets* (New York, 1977).

4. For a general overview of corporate involvement in electoral politics, see Edwin Epstein, *The Corporation in American Politics* (Englewood Cliffs, N.J., 1969); and "Corporations and Labor Unions in Electoral Politics," *Annals of the American Academy of Political and Social Science* (May 1976), pp. 33-58.

5. For a general political history of Texas in the twentieth century, see Seth McKay and Odie Faulk, *Texas after Spindletop* (Austin, 1965).

6. For a general history of the API, see Leonard Fanning, *The American Petroleum Institute* (New York, 1959). For a specific instance of API and government cooperation, see J. H. Nelson (Petroleum Section, Bureau of Foreign and Domestic Commerce) to L. Walker (API Division of Statistics), letter dated March 9, 1928, in Box 1653, Record Group 312, General Correspondence of the Bureau of Foreign and Domestic Commerce, National Archives. Much general information regarding the API's political activity in the 1920s is found in the Sun Oil Papers, Series 21-A, Eleutherian Mills Historical Library, Wilmington, Delaware.

7. Theodore Lowi, *The End of Liberalism* (New York, 1969). See, also, Grant McConnell, *Private Power and American Democracy* (New York, 1966). Two works by Robert Engler describe the political activities of the largest oil companies. See *The Politics of Oil* (Chicago, 1961) and *The Brotherhood of Oil* (Chicago, 1977).

8. See, for example, Francis X. Sutton et al., *The American Business Creed* (New York, 1956). For a more recent summary of corporate attitudes toward the state, see Leonard Silk and David Vogel, *Ethics and Profits* (New York, 1976), pp. 42-101. In discussing the rhetoric of the period at the turn of the twentieth century, Samuel Hays argues that "antimonopoly correctly describes the ideology of the conservation movement, but fails to analyze its broader meaning." See Samuel Hays, *Conservation,* p. 262.

9. *Annual Report of the Comptroller of Public Accounts of the State of Texas,* Year Ending August 31, 1900 (Austin, 1900), pp. 107-124; Seth McKay, *Seven Decades of the Texas Constitution of 1876* (Lubbock, Texas, 1942), pp. 83, 213; and *Comptroller's Report, 1900,* pp. 123-124.

10. Such shifts complicate a regional analysis of political change. Those events which most affected the refining region were initially handled at the state level, but the continued growth of the oil companies pushed the focus of their political activities (and therefore of this chapter) to broader levels.

11. The best analysis of these laws, as they applied to petroleum is in John O. King, *Joseph Stephen Cullinan: A Study of Leadership in the Texas Petroleum Industry, 1897-1937* (Nashville, 1970), pp. 112-145. See, also, Jewel Lightfoot, *Anti-Trust Laws of Texas* (Austin, 1907); H. R. Seager and C. A. Gulick, Jr., *Trust and Corporation Problems* (New York, 1929); Tom Finty, Jr., *Anti-Trust Legislation in Texas: An Historical and Analytical Review of the Enactment and Administration of the Various Laws upon the Subject* (Dallas, 1916); and John R. Allison, "Survey of the Texas Antitrust Laws," *Antitrust Bulletin* XX (Summer 1975), pp. 215-308.

12. For the staffing of the Texas attorney general's office, see *Annual Report of the Comptroller of Public Accounts of the State of Texas for the Year Ending August 31, 1900* (Austin, 1900), p. 123. For the Bureau of Corporations' report on the petroleum industry, see U.S. Bureau of Corporations, *Report of the Commissioner of Corporations on the Petroleum Industry,* 2 vols. (Washington 1907).

13. For an excellent discussion of national attitudes toward antitrust, see Ellis Hawley, *The New Deal and the Problem of Monopoly* (Princeton, 1966). For regional attitudes, see, for

example, *Oil Investors' Journal,* December 15, 1902, p. 5; January 1, 1903, p. 11; and January 15, 1903, p. 1.

14. *Beaumont Daily Enterprise,* April 6, 1901, p. 1; and *Oil Investors' Journal,* January 1, 1903, p. 13.

15. See, for example, the editorial, "Shall Texas Surrender Her Oil to a Trust?" *Beaumont Daily Enterprise,* January 23, 1901.

16. *Oil Investors' Journal,* November 19, 1907, p. 12. For a somewhat confused account of Magnolia's problems with Texas antitrust laws in this period, see Charles Wallace, *Nine Lives: The Story of the Magnolia Companies and the Anti-Trust Laws* (Dallas, 1953).

17. History of the Sale of the Security and Navarro Properties," File 1-1434-3, Federal Trade Commission's Oklahoma Oil Investigation, Record Group 122, National Archives.

18. FTC to G. C. Greer (Magnolia), letter dated December 7, 1917, File 7336-185-1, p. 76, FTC's Oklahoma Oil Investigation.

19. *Port Arthur Herald,* May 4, 1901, p. 1; *Oil Investors' Journal,* December 5, 1907, pp. 15-17.

20. William Larimer Mellon, *Judge Mellon's Sons* (Pittsburgh, 1948), p. 269.

21. Although Ralph Andreano cites the importance of Texas laws in opening the Spindletop field to new companies, the thrust of this work and of the more general study, *The American Petroleum Industry,* is to diminish the importance of political factors, especially the dissolution of Standard, and to emphasize the importance of economic factors in changing the market structure of the oil industry from near-monopoly to oligopoly. My research suggests that in the southwestern fields, which were of crucial importance in the microeconomic changes that he describes, the state, not the national, political system was a prime determinant of the market structure of oil. See Ralph Andreano, "The Emergence of New Competition in the American Petroleum Industry before 1911," doctoral dissertation, Northwestern University, 1960.

22. Robert Lively, "The American System: A Review Article," *Business History Review* XXIX (March 1955), pp. 81-96. See, also, Thomas McCraw, "Regulation in America."

23. Notes on interview of James Autry (Texaco), July 16, 1906, Drawer 397, #3208, p. 95, Bureau of Corporation Investigation of the Petroleum Industry, Record Group 122, National Archives.

24. James Autry to Joseph Cullinan, letter dated May 12, 1904, "Business Correspondence, 1901-1912," folder 1904, Box 27, Papers of James Lockhart Autry, Rice University, Houston, Texas.

25. James Autry to Joseph Cullinan, letter dated May 28, 1904, "Business Correspondence, 1901-1912," folder 1904, Box 27, Papers of James Lockhart Autry.

26. John O. King, *Joseph Stephen Cullinan,* p. 172.

27. Jewel Lightfoot, *Anti-Trust Legislation in Texas* (Austin, 1907), pp. 51, 55.

28. See "The Texas Company" folder, entries under "1915" and "1917" in L. W. Kemp Collection in Barker Collection, University of Texas, Austin, Texas. See, also, *Oil and Gas Journal,* January 28, 1915, p. 14; and Marquis James, *The Texaco Story: The First Fifty Years, 1902-1952* (The Texas Company, 1953), p. 53.

29. Richard Hofstadter, "What Happened to the Antitrust Movement?" in Earl Cheit (ed.), *The Business Establishment* (New York, 1964), pp. 113-151.

30. Louis Galambos, *The Public Image of Big Business in America, 1880-1940* (Baltimore, 1975).

31. The state of Texas planned ouster suits against Gulf, Texaco, and Humble in 1923. In 1933, Texas brought the large companies to court for antitrust violations. See *Oil and Gas Journal,* September 6, 1923, pp. 53-54. The TNEC Hearings in the late 1930s were aimed at tracing the extent of concentration in oil and other industries. Just before World

War II, the national government filed an all-embracing antitrust suit versus the major oil companies. See Gerald Nash, *United States Oil Policy* (Pittsburgh, 1968), pp. 154–155. Bills introduced in the 1970s ran from divestiture to constraints on further diversification into related energy fields.

32. For a discussion of the recent debates on antitrust in oil, see *Johns Hopkins University Conference on Divestiture, Capitalism, and Competition: Oil Industry Divestiture and the Public Interest* (Baltimore, 1976). For a strong advocate of antitrust, see John Blair, *The Control of Oil* (New York, 1976).

33. Walter Galenson, *The CIO Challenge to the AFL* (Cambridge, Mass., 1960).

34. So important was the total impact of the NLRB's activities in encouraging the expansion of the CIO that Walter Galenson—perhaps the leading authority on the labor disputes of the late 1930s—has argued that without its support, the nucleus of unionism established by the CIO in heavy industries could not have survived the depression years after 1938. This was a regulatory agency that mattered.

35. For a discussion of the NLRB's decisions in specific refineries, see Harvey O'Connor, *History of the Oil Workers International Union (CIO)* (Denver, 1950).

36. *Fortune,* October 1938, p. 18.

37. For an overview of government regulation of personnel practices, see Murray Weidenbaum, *Business, Government, and the Public* (Englewood Cliffs, N.J., 1977), pp. 92–117.

38. J. Clarence Davies III and Barbara Davies, *The Politics of Pollution,* 2nd ed. (Indianapolis, 1975); David Davis, *Energy Politics,* 2nd ed. (New York, 1977); and National Research Council, *Implications of Environmental Regulations for Energy Production and Consumption* (Washington, 1977).

39. Interview of George Craig, on file in Port Arthur Chamber of Commerce Historical Files (no date), p. 5.

40. This lobbying is most evident from letters throughout the period 1901–1906 in the Papers of George Craig, San Jacinto Monument Historical Museum, Deer Park, Texas; Marilyn McAdams Sibley, *The Port of Houston: A History* (Austin, 1968), pp. 135, 152, 170. For Beaumont, see Lois Hunt, "A History of the Beaumont Municipal Port, 1856–1942," master's thesis, University of Texas—Austin, 1943. For Port Arthur, see John Rochelle, "Port Arthur: A History of Its Port to 1963," master's thesis, Lamar College, 1969. Specific dates of improvements for each port are given in the Corps of Engineers Port Series.

41. For Cullinan's role, see "Intracoastal Canal Association, 1917–1922" folder. Papers of Joseph Stephen Cullinan, Gulf Coast Historical Association Archives, Houston Metropolitan Research Center, Houston Public Library.

42. For an excellent history of the Corps of Engineers activities on the Texas Gulf Coast, see Lynn Alperin, *Custodians of the Coast: History of the United States Army Engineers at Galveston* (Galveston, 1977).

43. F. T. Manley, "Material and Equipment in the Port Arthur Works," *The Texaco Star,* vol. IV, no. 7 (May 1917), p. 7; see, also, Louis Otts, Jr., "Water Requirements of the Petroleum Refining Industry," Geological Survey Water-Supply Paper, 1330-6 (Washington, 1963), p. 287.

44. *Beaumont Enterprise,* October 10, 1954, Chemical Edition, Section II, p. 2; November 5, 1971, p. 9.

45. Joseph Cullinan to Arnold Schlaet, letter dated April 26, 1904, Volume 29, *Texas Company Archives,* p. 155.

46. Gerald Nash, *United States Oil Policy,* (Pittsburgh, 1968), pp. 113–120; and Robert Hardwicke, *Legal History of Gas and Oil* (Chicago, 1938), pp. 214–220.

47. Robert Freeman Smith, *The United States and Revolutionary Nationalism in Mexico,*

1916-1932 (Chicago, 1972); and in Howard F. Cline *United States and Mexico* (New York, 1963), part III. See, also, Lorenzo Meyer, *Mexico and the United States in the Oil Controversy, 1917-1942* (translated by Muriel Vasconcellos) (Austin, 1977).

48. H. K. V. Tompkins to Arthur C. Frost, July 11, 1927; Frost to Secretary of State, August 11, 1927, Confidential Correspondence, Tampico 1924-1932, Records of the Consular Files of the State Department, National Archives, Washington, D.C.

49. Swain to Gunther, December 14, 1925, Document 812.6363/1636, Records of State Department, Record Group 59, National Archives, Washington, D.C. Before 1876, Mexico followed the Hispanic legal tradition in ascribing subsoil rights to the state. During the regime of Portfirio Díaz (1876-1910), leases granting subsoil rights to individuals were written in an attempt to attract foreign investment. The Mexican Constitution of 1917 reclaimed subsoil rights for the government.

50. Fletcher to Watriss, August 31, 1921, *Foreign Relations of the United States*, 1924, vol. II, p. 451.

51. For nonrecognition, see Eugene Traui, "Harding's Recognition of Mexico," *Ohio History* LXXV (Spring-Summer 1966), pp. 137-148. The arms embargo is discussed in Fletcher to Summerlin, April 11, 1919, Foreign Relations, 1919, vol. II, p. 594. Battleships were sent on several occasions. See, especially, Secretary of the Navy to Secretary of State, July 2, 1921, State Department 812.00/25070. For the diplomatic protest, see Summerlin to Secretary of State, November 17, 1922, Foreign Relations, 1922, vol. II, p. 203.

52. Samuel Flagg Bemis, *The Latin American Policy of the United States* (New York: Harcourt, Brace and Co., 1943), p. 217, cites this vote "as the really revealing feature of the Mexican policy, indeed of the Latin American policy of the United States." It was the symbolic end of intervention. Also, see the *New York Times,* January 22 and January 30, 1927, p. 1.

53. See Jerry Israel, "A Diplomatic Machine: Scientific Management in the Department of State," in Jerry Israel (ed.), *Building the Organizational Society* (New York: Free Press, 1972), pp. 183-196. See, also, John P. Harrison, "The Archives of the U.S. Diplomatic and Consular Posts in Latin America," *Hispanic American Review* 33 (February 1953), pp. 168-183; and Waldo H. Heinrichs, Jr., "Bureaucracy and Professionalism in the Development of American Career Diplomacy," in John Braeman (ed.), *Twentieth-Century American Foreign Policy* (Ohio State University Press, 1971), pp. 119-206.

54. Ralph W. Hidy and Muriel E. Hidy, *Pioneering in Big Business, 1882-1911* (New York, 1955), p. 135.

55. Morrow to Olds, letter dated May 8, 1928, State Department 812.6363/2563 1/2.

56. This process was especially difficult in the increasingly important Middle East, where the large American companies originally entered several countries before the State Department. See George Gibb and Evelyn Knowlton, *The Resurgent Years, 1911-1927* (New York, 1956), pp. 278-317; and Anthony Sampson, *The Seven Sisters* (New York, 1975).

57. *Oil and Gas Journal,* December 4, 1913, p. 2; February 12, 1914, p. 34; and February 5, 1914, p. 30. The best source of detailed information regarding Daniels' scheme is File 1430-1, Oklahoma Oil Investigation, Records of the Federal Trade Commission, Record Group 122, National Archives, Washington, D.C.

Chapter VIII

Institutional Responses To Refinery-Related Pollution

The growing refining complex on the Gulf Coast caused considerable pollution, but for most of the century no institutions—either public or private—had sufficient power or incentive to deal effectively with the resulting problems. Although both air and water pollution were widespread in the early years, they were generally not treated as major concerns. Beginning in the 1920s, a series of "crises" brought about by especially severe water pollution forced existing institutions to confront this issue. Inspired by the search for greater efficiency and by the desire to prevent politically imposed solutions, the large oil companies were the first to respond, but they experienced great difficulty in devising and sustaining organizational responses to pollution that could keep pace with increasing environmental problems generated by their pursuit of their primary organization goal, the production of more and better petroleum products. Their actions delayed government intervention, but even those public agencies that were potentially capable of attacking pollution faced institutional constraints somewhat similar to those that hampered oil industry responses, since pollution control was not the primary organizational concern of any of these public agencies. The persistance of water pollution and the additional impetus provided by air pollution problems in the 1950s finally forced the public sector to fashion agencies capable of addressing these increasingly serious concerns. The long neglect of environmental problems thus reflected the difficulties of creating new institutional arrangements capable of dealing with the externalities of growth. Existing public and private institutions only haltingly acknowledged and responded to changed conditions, and the public sector proved unwilling or unable to assert new authority in an area where decisions had traditionally been made by private corporations. In light of the lack of vocal, sustained pressures for pollution control from a society that was most reluctant to modify its commitment to rapid development, such substantial institutional barriers were long sufficient to block efforts by small groups in the society that sought stronger initiatives against pollution.[1]

The environmental impact of the growth of refining is best studied as one part of the broader history of the region's oil-related economic development. The three general eras in the history of regional refining identified earlier provide a useful framework within which to examine petroleum-related pollution problems: (1) the early years (1901–1916) of unrestrained production that resulted in localized pollution in the Port Arthur area; (2) a middle period (from about 1916 to the late 1930s) in which the geographical spread of refining and the introduction of new production techniques aimed at increasing the yield of gasoline brought more severe problems; and (3) a modern era (1941 to the 1960s) marked by the diversification of the basic causes of pollution due to the expansion of other major industries and the increasing use of the automobile. In all three periods, petroleum-related pollution was substantial. Despite the improved efficiency of the refineries over time, the absolute quantity of pollutants increased as the refining complex continued to expand and the chemical complexity of pollutants increased with the implementation of new production techniques.

Each of the three eras witnessed a fairly distinct pattern of response to pollution. During the early years, neither the state nor the oil companies made serious attempts to control the escape of oil into the region's air and water. In the middle period, however, the companies and their primary trade association, the American Petroleum Institute (API), took the lead in responding to petroleum-related pollution. They did this by closing up the production process to eliminate waste and by attempting to control the political process to forestall public intervention. Their efforts ultimately failed and, in the modern era, first state and local governments and then the federal government gradually took greater control in attempting to deal with the environmental problems brought about by the cumulative impact of more than fifty years of refining. Several significant changes in the responses to pollution over time are thus evident. Both the private and public institutions most directly concerned with petroleum-related pollution became increasingly national rather than regional. A second substantial shift was the gradual transfer of leadership in defining solutions from private to public institutions.[2]

THE ERA OF THE GUSHER (1901–1916)

Before depicting the environmental impact of the petroleum industry on the upper Texas Gulf Coast, a brief reminder is in order. The discovery of oil did not mark the beginning of the region's problems with pollution. Earlier, coal-burning trains had brought unsightly and unhealthy smoke, a hazard that was especially bad in rail centers like Houston. The rapid cutting of timber had stripped the soil of its protective covering in parts of East Texas; the milling of this lumber in coastal towns like Beaumont had resulted in contamination of the air. Cotton and timber production had been rapidly expanded without regard for the effects on the soil or the forests. Municipal waste disposal was also a major source of water pollution. In 1913—before the construction of large refineries in the Houston area—the Mayor of Houston acknowledged that sewage disposal had caused local pollution that was "little less than disgraceful."[3] In the long run, the combined influence of such abuses would have led to extensive damage to the region's land, air, and water systems, even without the expansion of the oil industry.

Nevertheless, the discovery of oil brought significant new types of pollution, greatly expanding the region's environmental problems. It brought an era of much more rapid development and, in petroleum, a new and more troublesome source of pollution. In the rush for instant wealth, oil was seen as black gold, not black sludge. The gusher was the ecological symbol for this early period—photographs of the Lucas gusher went out across the world, showing the magnificent spectacle of a six-inch stream of oil rising more than 100 feet above the top of the derrick. So powerful was this image that wise well owners arranged to turn on similar gushers for the entertainment and persuasion of potential investors.[4] In this atmosphere of uncontrolled exploitation, few cameras recorded the fate of the gusher-produced oil after it splashed to earth.

The gushers were only the most visible sources of pollution. In this formative period, a regional oil industry was constructed from scratch. In an all-out race to beat competitors to market, new facilities were generally hastily built. Speed and quantity of production—not efficiency and the prevention of waste—thus characterized this early period. As a result, the race to market was run over a very slick track. Flush production led to often massive drain-offs of crude that soaked the ground; the rapid removal of oil from the wells also resulted in the introduction of salt water into the underground reservoirs of oil and into the region's water system. For lack of better storage, producers often used open earthen pits and wooden tanks. Adequate transportation and loading facilities did not exist, so large quantities of oil were lost between pumping stations and the tankers that carried much of the crude to the East Coast. It is likely that at least as much oil found its way into the region's ground, water, and air in this period as found its way to market.

Inside the early refineries, all aspects of the operations reflected a general lack of regard for the environmental consequences of unrestrained exploitation. Throughout much of the period up to 1914, numerous topping plants attempted to survive by distilling off only the most valuable products, like kerosene and gasoline, and disposing of the remainder of the crude. The large plants owned by Gulf, Texaco, and Magnolia (now Mobil) recovered a greater portion of the total amount of crude they processed, but they too relied on the simple technology of straight distillation in stills heated by burners fueled by crude oil. Such open flames were economically and ecologically inefficient, using large amounts of oil while producing a heavy smoke. Sulphur fumes created during refining escaped into the air, as did hydrocarbon vapors from poorly constructed storage and transportation equipment. Widespread water pollution resulted from frequent oil spills at the plants and at refinery shipping terminals. Flood control equipment was inadequate, and periodic floods in the coastal region washed much of the spilled oil into rivers and streams. Methods for separating oil out of the water used in refining operations were rudimentary and did not prevent the return of much oil-polluted water to the regional water system. Oil particles thus entered the region's lakes and, ultimately, the Gulf of Mexico. Here they joined large amounts of oil lost during the loading of tankers and contaminated ballast dumped at sea by the oil-burning tankers that were replacing coal-burners. By any measure, past or present, these rudimentary refineries and their supporting distribution systems were major sources of water and air contamination.

In the early period, oil pollution of air and water was both extensive and obvious in the Port Arthur-Beaumont area. Although the exact levels of such pollution can only be roughly estimated, it is clear that air pollution was a problem at a very early date.[5] Several days after Spindletop came in, a thick, yellow, sulphur-laden fog began periodically to discolor Beaumont houses. Similar fumes threatened the lives of some of the drillers. With the decline of crude production, the region's air quality improved (or, more precisely, the level of visible air contamination decreased), and the large refineries then became the primary source of air pollution in the Jefferson County area. Because of the distance between them and the surrounding cities, and because of prevailing ocean breezes which removed contaminated air, they did not at first cause a major air pollution problem.[6]

Such was not the case for water pollution, however. Fourteen years of increasing discharge led to serious levels of contamination in the regional waterways. Like refinery-related air pollution, this problem was concentrated in the Port Arthur area. But it also spread outward along the paths taken by the oil-carrying and oil-burning tankers, and conditions in the shipping lanes and at the loading and unloading terminals on both ends of the voyage reflected those around Port Arthur. Both water and air pollution, however, went largely unnoticed for most of the early period.

The blatant disregard for the waste of large quantities of oil during the era of the gusher seems very remote in the current age of energy shortages. Nevertheless, many of the attitudes toward pollution from the era of unrestrained expansion proved extremely tenacious, and many of the institutional problems that then stood in the way of attempts at pollution control continued to hamper the search for solutions. An examination of the attitudes and institutions of the formative years of the region's oil industry thus suggests several constraints on pollution control that continued to hamper antipollution efforts for much of the remainder of the century.

In this early period, the primary large-scale organizations in the region, the oil companies, had no incentive to recognize and respond to the pollution problems caused by their operations. Crude prices dipped to as low as three cents per barrel, making the spillage of a few barrels, or even a few hundred barrels, of minor economic importance to the owner. A highly competitive market structure made short-run survival more crucial than long-run planning, and the low level of available technology led to the recovery for sale of only a small portion of each barrel of crude. Because of the small size and limited resources of many companies, expenditures for control of pollution were an unaffordable luxury, at least from a microeconomic perspective. When viewed in the context of a larger society that made few demands for the control of pollution, this pattern of business behavior becomes all the more understandable; the situation in this early period encouraged speed and quantity of production, not efficiency and the avoidance of waste. The oil companies did as the market they faced dictated: they chose rapid, unrestrained production with minimal concern for the escape into the air and water of large quantities of their not-so-valuable raw material.[7]

The dominant values of the region and of the larger society of which it was a part further discouraged concern about pollution. Growth was the primary economic goal shared by most segments of society. Restraints on rapid growth and impingements on the rights of property owners would have been required to deal with the oil-related damage to the environment. Neither was likely to occur until oil pollution reached crisis proportions that threatened other industries or the health of those in the region. In the absence of such a crisis, the agents of rapid growth—the oil companies—enjoyed great leeway in determining what was best for their own well-being and, by extension, for that of the region.

This early period of nearly unchallenged oil industry dominance had a long-lasting influence on government-business relations. Through an ideological lens of "free competition," oil executives in the Spindletop era viewed government suspiciously, as a potential usurper of corporate power and a threat to corporate autonomy. Politicians were usually seen as meddlers who were both opportunistic and incompetent. The legislation they produced was, almost by definition, confusing and wrongheaded. Extremely threatening campaign rhetoric by office seekers and periodic antitrust assaults on the oil companies reinforced these perceptions. The resulting distrust of and disdain for government did not disappear. Of course, such attitudes did not prevent business from cooperating with government on measures that were beneficial to it. But when government attempted to assert power in areas previously controlled solely by the corporation, cooperation became most difficult. Pollution control was one of the most volatile such issues.[8]

The oil companies' ideological devotion to free enterprise thus reinforced a very practical aversion to sharing power. But the oil executives' mistrust of government officials was not simply the result of their own particularly strong ideological and practical disdain for government. Businessmen often perceived much government activity as incompentent and misguided because it was, in fact, incompetent and misguided. In the early period and throughout much of the century, various levels of government were often simply not capable of governing; they could not effectively respond to the wide variety of demands placed on them by rapid growth.[9] Such deficiencies in the public sector did not arise entirely because politicians were opportunistic or poorly suited for their offices. Indeed, sufficient structural problems existed within the public sector to account for many of its shortcomings.

Because of the lack of better solutions, several state agencies became, by default, the focus of pollution-control efforts for almost sixty years. From the beginning of the region's oil-induced development, these agencies were ill-equipped to deal with the problems that accompanied the sudden intrusion of the oil industry. The state government lacked the money, personnel, and expertise to do much more than acknowledge that pollution existed. The Fish and Oyster Commission, which became one of the agencies responsible for controlling water pollution, consisted of just one man with annual expenses of $2,400. The State Board of Health, which later took on an important role in pollution control, had substantially greater resources, but most of its expenditures involved the inspection of incoming ships.[10] The resources of state agencies expanded gradually throughout the century, but they were still meager weapons with which to combat the pollution caused by one of the world's largest refining complexes. From 1901 until the 1960s, these state agencies—underfunded and understaffed as they were—were the center of public initiatives to control oil pollution.[11]

Perhaps even more significant than the general lack of resources was the absence in the early years, as throughout most of the century, of even an underfunded public body with a specific and overriding institutional mandate to control pollution. Instead, each of the existing agencies reacted to pollution as a minor problem that was largely incidental to its central duties. Thus, the attorney general's office prosecuted nuisance cases against polluters, but it focused most of its energies on other types of cases. Similarly, although pollution sometimes became a threat to public health, local and state public health departments had many more pressing demands on their resources. As the scale and complexity of petroleum-related pollution mounted, such inadequacies in the existing institutional arrangements in the public sector became increasingly evident.

Such problems were to be expected when a political structure inherited from a much less complex era attempted to respond to difficult problems brought about by the very large production of a relatively new product by corporations operating on a scale previously unknown to the region. The growing oil industry and the general regional development induced by oil created pressures for new public services and regulations. In this sense, the rapidly expanding industry pulled the public sector into the twentieth century, encouraging the modernization of government at many levels.[12]

In the early period, the public sector had great difficulty in responding to these new demands, since it lacked the experience and resources of the dynamic private corporations. The evolution of an efficient and independent public sector capable of responding to the demands created by sustained industrial growth was a lengthy and often painful historical process. As the oil companies that controlled most of the region's refining capacity continued to expand, they remained more highly organized—geographically, technologically, and administratively—than the various levels of government under which they operated. This continuing lag in capabilites resulted in part from the public sector's late start and its confusion as to purpose and authority. But it also reflected the scale and complexity of the problems that accompanied growth and the variety of difficult demands that were thus made on the public sector.

A lack of efficiency went hand-in-hand with a lack of independence. The oil companies had a very strong interest in identifying and attempting to control political or legal threats to their autonomy, and they thus moved quickly into the political arena in response to the uncertainty introduced by potential environmental controls. In a political system that responded to organized, sustained, interest-group pressure, the large oil companies were in an excellent position to acquire government favors and to resist government challenges.

Their adeptness at political manuevering reflected both the oil companies' organizational strengths and structural characteristics of the political system that made public responses to problems like pollution most difficult. The twin pillars around which the system was built—federalism and checks and balances—insured that political power was dispersed. Existing political boundaries did not match the boundaries of the region's growing pollution problem. Political units at the local, county, state, and national levels were, at least theoretically, capable of exerting control over portions of the problem. In practice, however, an often debilitating confusion of authority resulted, especially in dealing with a pollution problem that resolutely refused to be turned back at the county or state line and, instead, continued to grow across political boundaries.[13] Even when authority over a portion of the problem could be asserted by an existing agency, its limited resources and restricted time horizon generally prevented a forceful, systematic response. Thus, amid an overabundance of potential public authority, there was a paucity of institutions capable of exerting real power in the area of pollution control.

In the early period, those agencies that might have attempted to control petroleum-related pollution lacked more than resources; they did not even possess the clear legal authority to forbid the dumping of pollutants into the air and water. During the first twenty years of regional development, no local or state laws prohibited such discharge. The existing national law, the Rivers and Harbors Act of 1899, also provided little practical means for the control of oil pollution. The law had been framed to meet conditions prevalent when "the production and use of petroleum oil was insignificant," and it was "not considered broad enough in its meaning or character" to include most petroleum-related pollution.[14]

Such limited authority, combined with equally limited incentives for action, debilitating structural problems in the public sector, and a societal commitment to rapid growth, meant that little would be done about regional pollution until it reached crisis proportions.

THE ERA OF INDUSTRY LEADERSHIP (1916–1941)

Such a crisis came in the early years of the 1920s. The extraordinary water contamination that accompanied the rapid expansion of the region's oil industry during and immediately after World War I focused regional and national concern on water pollution. In response, during the 1920s and 1930s there were several significant departures in public regulation by the state and national governments. These public initiatives, however, were constrained by many of the same attitudinal and institutional problems that had previously blocked governmental action. In this middle period (1916–1941), the oil industry, not public agencies, generally played the dominant role by implementing private pollution control measures while striving to contain public measures that threatened corporate control of environmental affairs.

A close examination of the responses of the oil companies and the state to the water pollution crisis of the early 1920s provides useful insights into the relative power of each. Indeed, as the first major response to water pollution by oil in the twentieth century, the events leading up to the passage of several oil pollution laws in the 1920s set the tone for much of the period until the 1960s.

Before the crisis of the early 1920s could be treated, it had to be identified and measured. In response to public outcries and to complaints by other industries being affected by petroleum pollution, the API, the American Steamship Owners' Association, and the U.S. Bureau of Mines cooperated in completing a comprehensive investigation of national conditions. Compiled by representatives of the government and of the large oil companies and published in 1923, *Pollution by Oil of the Coast Waters of the United States* painted a bleak portrait of a national whose coastal waters were covered with oil. This descriptive report provided the first general survey of regional pollution, and it serves as a convenient benchmark for discussion of the levels of regional water pollution after 1923.[15]

After studying ports along the East and Gulf Coasts, the investigators concluded that Port Arthur "with its exceedingly heavy oil commerce, represents one of the worst, if not the worst, condition of oil pollution witnessed by the Committee." So bad were the spillages in transferring oil to ships and the run-offs from refinery operations that "the large refiners maintain crews of men and equipment for skimming the surface of the harbor waters." Their efforts were not very successful, however: "they appear to be doing little more than holding their own." Beaumont and Galveston were somewhat cleaner. But the Houston Ship Channel "undoubtedly represents one of the worst oil polluted localities seen by the Committee" and was surpassed in this regard only by the channel at Port Arthur. Since the Houston waterway had been in use by oil companies for only seven years at the time of the investigation, its concontamination had been extremely rapid and thorough. The survey cited various sources of regional oil pollution, including dumping of oil at sea, run-offs from one oil field near Houston, and the careless handling of fuel oil by railroads and other industries. But in the Houston-Port Arthur areas, the primary sources were industrial waste and the escape of oil during transfer from refinery docks to seagoing vessels.[16]

The 1923 investigation suggests that the region experienced perhaps its worst surface water pollution of the twentieth century in the early 1920s. This contamination drove away the lumber industry, whose product was ruined by oil, from Port Arthur to Orange and Beaumont. Gulf Coast fishing and oyster dredging suffered. The wildlife that had initially helped to attract hunters and fishermen to the area in the 1890s became less plentiful. The region's beaches became nearly unfit for recreational use. At Galveston, where tourism was the major business, oil pollution in the water and on the beaches became "extremely bad," so bad, in fact, that cans of gasoline were placed along the beach so that bathers could remove the oil accumulated while swimming.[17]

Such conditions focused public attention on the inadequacies of the existing laws; several new laws resulted. In 1918, the Port of Houston passed a harbor regulation that prohibited the dumping of untreated oil products into the Houston Ship Channel. In 1924, the federal government passed the Federal Oil Pollution Act, which regulated the dumping of fuel at sea by oil-burning vessels.[18] These statutes were among the first aimed specifically at water pollution by oil, and they were important in establishing the legal basis for the control of all types of pollution.[19]

Significant as they were, these laws had only limited impact because of the way they were passed and enforced. The Port of Houston regulation explicitly prohibited the discharge of harmful oil products. But when the new port director, B. C. Allin, who had been recommended for the job by a former president of Humble Oil, attempted to enforce the law as he interpreted it, he faced heated opposition from various refiners. After failing to obtain court convictions in cases against numerous polluters—despite what he viewed as ample evidence—Allin called for city council measures to assure that "proper convictions" could be obtained. This action, and the belief that the director was harassing the industries along the ship channel while attempting to usurp jurisdiction that was not his, gave the companies an opening; the attorney for the Galena-Signal Oil Company demanded that the mayor of Houston replace the port director. Allin had antagonized influential oil men like the president of Galena-Signal, Joseph Cullinan, who had been president of Texaco until 1913. Cullinan, along with other founders of the regional oil industry, had given much of his time and a considerable amount of his own money in advancing the interests of Houston and its waterway before Congress. In fact, it had been Cullinan who had originally suggested that the city of Houston exercise police powers along the ship channel, but he was now angered when these powers were used against his own firm and others in that area. Important oil men considered Allin's actions as a way of using "trifling matters to harass companies which had invested millions in plants along the waterway." They were utilizing, they felt, "the most modern appliances to prevent the escape of oil." Against such opposition, an appointed public official whose primary duty was to oversee the rapid development of the port was not in a good position to demand further water treatment equipment. Subsequent actions aimed at curbing oil pollution were more restrained.[20]

The oil companies exerted similar influence on the national law passed in 1924, although in this case their efforts were less direct. Protests by other industries adversely affected by petroleum pollution and by sportsmen created much of the demand for legislation, but the oil industry proved very adept at controlling the actual content of the act that finally passed. The API-Bureau of Mines report discussed above served as an important source of information for those who framed the Federal Oil Pollution Act of 1924. After providing much of the information available to Congress, the API and the large oil companies that were its major members used their political power to constrain the scope of the law. In the end, oil industry interests succeeded in limiting the law to coverage of oil discharged by vessels at sea, despite the fact that even the API-Bureau of Mines report had also cited refineries and oil field runoffs as important sources of pollution.[21]

A similar political strategy of containment also succeeded in the late 1930s, when water pollution control measures received an unprecendented amount of attention in the national legislature, and, for a time, a national stream pollution control act seemed probable.[22] A study prepared for the API on the bills pending before Congress was unusually explicit in suggesting how such bills could be influenced. The study's recommendations to the API were in the form of a comprehensive strategy for controlling the fate of pollution legislation. It stressed that the Texas delegation in Congress, which included the chairmen of several important committees, was of "the greatest possible strategic value." Thus, in the Texas delegation and in those of other southern and southwestern states, the oil industry's "not inconsiderable influence can be quietly exerted in any number of ways . . . setting in motion throughout the Southern states those forces of opinion to which members of Congress are especially sensitive." Special news features could be used to educate the polity, but "public statements should be confined to groups of producers—preferably independent organizations which are free of the political liabilities that some of the industry's major units have inherited from a fairly recent past." The report suggested that industrywide lobbying and possibly the cooperation of other affected industries should be mobilized in support of state and interstate action on water pollution in order to decrease pressures for federal legislation. Above all, the industry should not compromise by recognizing the principle of federal regulation, since "the only thing which practical politicians . . . recognize is an organized opposition capable of influencing votes. If the industry succeeds on that basis—plays poker rather than throwing down its cards in advance—it may again, as it did in 1924, postpone drastic anti-pollution legislation."[23]

This assessment proved unduly pessimistic; almost thirty more years were to pass before the enactment of a strong national water pollution control law in 1965. The key to the oil industry's ability to control the political environment was its access to the decision makers in Congress and the lack of equal access by those with different interests. The relative balance of political power between polluters and those affected by pollution was correctly perceived in industry discussions regarding the possibility of legislation: "Action is likely, not to appease sportsmen or Walton leagues (which are largely without political influence), but because of purely political factors."[24] In a political system characterized by dispersed power, those who could identify the crucial centers of power and bring political pressures to bear on those centers could generally succeed in attaining their political interests. The individual oil companies and the oil industry in general were well-suited for success in a broker state political bargaining process; those groups that gradually emerged to challenge industry lobbying efforts in the area of pollution control were not so well-suited. They lacked the clarity of purpose, the unity of goals, the scale of political resources, and the political experience of the oil industry interest groups. As a result, in the 1930s, and for several decades thereafter, effective national legislation aimed at curbing petroleum-related pollution never quite made it through Congress, despite widespread rhetorical support for such a measure.

Water pollution thus remained primarily a matter for state concern. The experience of Jefferson County (Beaumont and Port Arthur) with oil pollution in 1939 provides a good example of the difficulties of handling such problems at the state level. When salt water from producing wells in East Texas flowed down the Neches River and threatened Jefferson County's fresh water supply for domestic and industrial use, local authorities had no established means to remedy the situation. There was still no single state agency whose primary task it was to deal with such matters. Given this institutional-jurisdictional void, four agencies that were peripherally concerned with such problems issued a joint warning: the action was taken by the Texas Railroad Commission, which regulated oil production; the Attorney General's office; the state Board of Health; and the Game, Fish, and Oyster Commission. The threat of legal action led "several of the larger oil operators" to make improvements. But substantial pollution continued, and it led to a suit seeking an injunction against the producers who were allegedly causing "irreparable injury to property rights and civil rights, for which no adequate remedy at law exists." Although hailed as a pioneer case, the suit was compromised by the Texas attorney general. The Beaumont city attorney and the Jefferson County group that had pursued the matter for over a year and a half objected, but to no avail. In return for dropping the suit, the oil companies involved promised to stop polluting. They paid no penalties, however, for past abuses. Since the state attorney general was serving out a two-year term, his acceptance of "full responsibility" for enforcement was unconvincing to Beaumonters, who feared that the oil companies would not respect the agreement.[25]

Reliance on the promise of an elected official to enforce a voluntary agreement pointed up the absence of an adequate institutional framework for dealing with the external, environmental costs of economic growth. Several state agencies had limited authority over aspects of pollution-related problems. In each case, however, pollution control was a secondary concern of an organization whose primary functions, allocated resources, and trained personnel dictated greater concern for other tasks. Lacking the administrative machinery to handle even the Jefferson County case, which involved obviously harmful pollution clearly traceable to a known source, the state of Texas was ill-equipped indeed to control the more complex air and water pollution that accompanied the subsequent development of the refining region.

PRIVATE RESPONSES IN THE MIDDLE PERIOD

During the middle period (1916–1941), only limited political initiatives thus were taken at either the national or state level, but the public debates over control seem to have had some impact on the search for private solutions. Throughout this period, the large refiners and the API pursued private measures aimed at ameliorating pollution. The incentive for such actions was primarily economic; efficiency meant higher profits. But political factors undoubtedly reinforced these economic considerations. The desire to contain and ultimately to control politically imposed solutions accelerated the pace of the oil companies' search for cleaner methods of production. An oil industry spokesman summarized this attitude in 1936. He noted that, in the past, "as a rule, it has been only under threats from authorities that action has been taken." This had resulted in much ill-will while increasing the demand for new laws. To be in a position to "demand a voice in creating the rules under which it must play," however, the industry must have "clean hands" and "a clean house."[26] The fear of government intervention was a strong secondary motive behind private responses to pollution in the 1920s and 1930s.

The API was in the forefront of these efforts. After aiding the Bureau of Mines in the investigation of oil pollution in 1923, the API conducted its own follow-up survey in 1927. Its report cited a "marked improvement in oil pollution conditions" in support of its contention that "existing legislation is adequate" and "that better results can be obtained by voluntary effort and cooperation than by further legislation." This survey was at once a public relations measure aimed at soothing public concern and a serious report on private initiatives in the area of pollution control. Subsequent efforts of the API also had this dual nature, but they should not therefore be categorized merely as "P.R." and dismissed as meaningless. The public relations value of such activity was grounded in the specific, industry-sponsored research programs which laid a foundation for understanding the impact and treatment of oil pollution. An industry spokesman in a later period aptly summarized the API's attitude in the middle period when he said that "the best defense the industry has against unnecessary corrective measures and an out-of-proportion control effort is a heavy backlog of definitive information."

Beginning in the 1920s, the API's research was a primary source of such information for the oil industry, for political institutions involved in regulating the industry, and for the general public.[27]

The API's first discovery was the void of knowledge concerning petroleum pollution, and it set out to begin to fill this void.[28] Its environmental research aimed at accumulating and compiling practical knowledge useful in building and operating cleaner refineries. One of the earliest results was the three-part *Manual of Disposal of Refinery Wastes,* which outlines the best methods for the disposal of waste waters containing oil, waste gases and vapors, and waste waters containing solutes.[29] The API's applied research focused on the development of a modern, efficient separator. The specifications and standards that the Institute established were accepted and used throughout the industry in coping with two related problems: the separation of oil from water prior to its discharge from the refinery and the treatment of trapped wastes.

Environmental research within the individual oil companies also helped to fill in gaps in the understanding of waste disposal. Atlantic Refining Company led the way with a "Division of Waste Control" that had its own separate laboratory by 1933.[30] This commitment of resources to environmental research marked a significant departure by establishing both research programs and research institutions which could be expanded in later years.

Unfortunately, such potentially significant initiatives had little impact on pollution control during the middle period. For while a small group of industry researchers found answers to the most basic questions involving the treatment of water discharge, a much greater industrywide research effort developed new techniques of production. Indeed, a defining characteristic of the middle period was the introduction of a series of important technological breakthroughs that resulted in increased yields of improved gasoline from each barrel of crude refined. Such advances, however, also brought increased yields of "improved"—that is, more chemically complex—types of pollution. Thus, while a few researchers developed an efficient separator to control water pollution, many more introduced changes in production technology that only much later would be recognized as important departures in the production of pollution.

This lag between implementation of production technology and the introduction of control technology can be seen most clearly in the case of tetraethyl lead. As refiners searched for ways to improve the quality of gasoline in the 1920s, the laboratories of General Motors made a major breakthrough by adding tetraethyl lead to gasoline. Thirty years later, researchers were still unraveling the basic chemistry of air pollution and were only beginning to understand the general nature of this phenomenon. Responses to the specific atmospheric effect of lead additives did not come until nearly forty years after their introduction. A similar lag in understanding was apparent—at least in retrospect—in the development of fluid cracking units in the late 1930s and the 1940s. At that time, refiners hailed the new unit as a major breakthrough in gasoline production. Only thirty years later, however, did environmental researchers begin to recognize and react to the fact that it also represented a major breakthrough in the production of air and water pollution. Other similar advances in production technology generally had long-run environmental implications that were not recognized, much less planned for, in the rush to produce more and better refined products. As a result, despite the industry's pioneering research efforts in pollution control in the middle period, the imbalance between production and control techniques increased as the refineries became more technologically advanced.[31]

Although the oil companies had little success in anticipating the environmental consequences of new technologies, they proved much more successful at eliminating the older problems of waste. The most significant antipollution measures taken by the individual refineries in the middle period were those directed at increased efficiency, and they were greatly encouraged by the rising value of petroleum. The *Oil and Gas Journal* cited the main rationale behind such measures in an article in 1926: "Reduction of waste is nothing more or less than increased efficiency in the management of the various departments of an oil company." The goal of waste reduction was increased profits, but an indirect and usually unnoticed consequence was often pollution reduction. In the early 1920s, for example, Texaco's Port Arthur works "undertook a campaign of fuel efficiency in order to reduce, if possible, the enormous quantity of fuel oil consumed under stills and boilers." "A very noticeable" reduction in fuel waste resulted, conserving oil and eliminating excessive smoke as well.[32] Even greater economic and environmental benefits accompanied a more radical change in fuel use, the substitution of natural gas and refinery gases for oil fuels in the 1920s and 1930s. Another important aspect of the campaign for increased efficiency involved the prevention of excessive evaporation from storage tanks; the construction and sale of improved tanks designed to cut down evaporation losses proliferated in the 1920s. While saving valuable oil, such tanks also prevented the discharge of potentially harmful hydrocarbon vapors.

The rising cost of oil was incentive enough for increased efficiency, but other, noneconomic factors often entered into such considerations:

> The installation of suitable drainage systems and separators in refining is primarily a measure to prevent industrial waste, although it seems probable that, in order to comply with the laws against oil pollution, many refineries have installed equipment much more extensive than the mere prevention of industrial wastes would appear to warrant.[33]

In such cases, the desire to save oil was reinforced by the desire to avoid intervention by the public sector; the results were a reduction in waste and a decline in water pollution.

Substantial long-run decreases in the volume of refinery discharges also resulted from the often ingenious transformation of waste materials into useful products. In the early period of regional development, most firms sold only one or two primary refined products and disposed of the remaining by-products. Gasoline, natural gas, sulphur, sludge, coke, and a variety of other commodities once discarded as waste later became important products. Beginning in 1932, for instance, the petroleum coke that had long accumulated at Port Arthur area refineries was sold to Great Lakes Carbon Company, which manufactured pure carbon from it. Later, recovery units began to convert sulphur fumes once released into the air and spent acid once dumped into rivers into marketable sulphur and fertilizer. After treatment, sludge acid and other residues became fuel, as did the carbon dioxide derived from the carbon monoxide fumes created in the recovery of catalyst material. As a result of such efforts, the level of pollutants released in refining declined as those who could afford new processes, primarily the large oil companies, found ways to salvage useful products out of wastes.[34]

These measures were part of a continuing search for methods to extract greater value from each barrel of crude refined. The firms tried to minimize losses of oil during processing, increase the yield of the most desirable end products through sophisticated cracking techniques, and find markets or industrial uses within the refineries for as many by-products as possible. These trends characterized regional refining throughout its history, but economic, competitive pressures among large refiners accelerated the drive for efficiency during the middle period, just as political pressures to control environmental contamination were to do in the 1960s. The companies profited from the expansion of their marketable goods; the regional society profited from the reduction of discarded pollutants.

Such on-going reductions in the amount of pollutants discharged per barrel at least partially offset the increased absolute level of pollution caused by the sustained growth of refining throughout the upper Texas Gulf Coast. Although still a regional problem, the severe surface water pollution of the early 1920s probably declined in the next two decades.[35] New, more difficult problems accumulated, however, as a result of the introduction of new production technologies. In particular, there was little awareness of the coming problems with air pollution. Thus, although the oil industry took the lead during the middle years in combating the most flagrant forms of water pollution, other more complex environmental problems that were largely ignored in this period would later call forth stronger public initiatives.

THE GRADUAL EMERGENCE OF PUBLIC LEADERSHIP (1940–1960s)

World War II and its aftermath brought a period of extraordinary expansion to the upper Texas Gulf Coast. The continued growth of the petroleum industry and the rapid rise of several major new regional industries—notably petrochemicals and primary metals production—fed a post-war surge of economic development. The resulting industrial waste—combined with the added municipal waste caused by a growing population and the increased air pollution that accompanied the expanding use of the automobile—created major new environmental problems.[36] In the 1950s and 1960s, in the face of increasing air and water contamination from the cumulative impact of a half century of growth in population and industry, those in the refining region sought a better understanding of the causes of pollution and more forceful means of implementing solutions. The means chosen were usually in the public, not the private, sector. The large refiners' historical role as the primary source of much of the area's pollution meant that, no matter how well their record compared to that of other industries, they faced stricter public scrutiny in these years.

The post–World War II era witnessed continued problems with oil-related water pollution, but the "new" problems associated with air pollution absorbed much of the region's control efforts. By the mid-1950s, air contamination could simply no longer be ignored. The smog problem of Los Angeles had captured the nation's attention in the late 1940s. In terms of its expanse and its freeway system, Houston resembled Los Angeles, and it developed a similar air pollution problem. One response was a survey of air pollution in the years 1956–1958. The study's conclusions suggested the intensity of the problem. It recorded "localized," not "communitywide," air pollution. Although the photochemical reactions that contributed to Los Angeles' smog were not in evidence, the survey reported a variety of chemical pollutants from automobiles and from industry, including hydrocarbons in the ship channel area, where "approximately equal concentrations result from oil refining and other industrial operations."

The Beaumont-Port Arthur area, which had far fewer automobiles, did not experience such serious air contamination. Beginning in the late 1940s and 1950s, however, it did confront air pollution problems of primarily industrial origin. Oil and chemical fumes, believed to be potentially hazardous sulphur compounds, caused great discomfort, discoloration of houses, and bad odors. No survey equivalent to Houston's report in the 1950s recorded the precise level of air pollution in the Beaumont-Port Arthur area. But there and throughout the refining region in the late 1950s, widespread air contamination joined the water pollution inherited from an earlier era to create often severe problems for many in the refining region.[37]

Those most apt to seek new public controls over this pollution were those most directly exposed to it. In the early 1950s, a group of citizens who lived near the Houston Ship Channel began a private campaign to reduce pollution through meetings with local industries. They sought to extract voluntary promises under the threat of court action; there was no public agency other than the courts to which they could turn. Their protests generated pressure for the creation of such an agency and encouraged the formation of an air and water pollution control section in the Harris County Health Department (1953). Despite limited resources and its status as a secondary group within a larger organization with diverse functions, the new section utilized publicity, private conferences, and court action to make some progress in controlling water pollution. In the late 1950s, however, two developments limited its power to force reductions in discharges. First, a series of state court decisions made it much more difficult to prosecute offending corporations. Then, in 1961, the state of Texas established a state Water Pollution Control Board, whose jurisdiction conflicted with that of the Harris County agency.[38]

The organization of the state board shifted the focus of political action from a highly industrial, highly polluted region to the state legislature of a large, relatively unpolluted state—thereby undermining the campaign to achieve more effective governmental solutions. The traditional apportionment of state representation meant that agricultural sections held greater political power than the rapidly growing urban-industrial centers. In addition, the support of powerful oil producers from throughout the state could be mobilized to block moves against the coastal refiners. The evolution of the state board and its policies showed the effects of these political factors. Its program included the use of "grandfather clauses" that permitted continued pollution by existing industries, and it freely used its power to grant waste disposal permits. Even the charter of this first state pollution control agency reflected its weakness; included was a clause requiring behavior consistent with the "industrial development of the state." All in all, in its early years, the board did not hamper the growth of industry or of pollution.[39]

By the mid-1960s, such inaction was a luxury the region could no longer afford. In 1965, a follow-up survey to the Houston air pollution investigation of 1956–1958 revealed how quickly the limits to natural pollution dispersal could be reached in a period of rapid growth: "The most significant change which has taken place is the occurrence of the type of chemical reactions in the atmosphere which are characteristic of the well-known Los Angeles smog situation."[40] Water pollution also demanded attention. In 1967, federal investigators found that "the Houston Ship Channel was by far the worst example of water pollution observed by the board on its Texas tour." The problem was "overwhelming," and local officials were reported to feel that "it's so bad already, we don't need to worry about adding more pollution."[41] The investigators cited industry as the biggest polluter, with municipal sewage as a major secondary source. Despite forty-four years of gradual experimentation with water pollution control devices and laws, the 1967 report thus sounded disturbingly similar to that of 1923. Such conditions generated strong pressures for solutions, and Texas state authorities were soon forced to respond.

A newly activist national government further encouraged state action; under the threat of national legislation (in the absence of stronger state antipollution efforts), Texas passed a clean air act in 1965, followed by a water quality act in 1967.[42] These two measures gave public institutions at the state level the potential to exercise much stricter control of air and water pollution. For oil-related problems, this potential could be realized only when the agencies matured into regulatory bodies independent of oil industry control. But at least now the necessary institutions for implementing effective government control existed.

The early history of the implementation of these acts did not indicate that the agencies involved were going to take strong, independent action. Both laws called for programs consistent with "the operation of existing industries and the economic development of the state." Most of the original nine members named to the board established for the enforcement of the clean air act seemed to be citizens who would take this admonition seriously. Public protests followed the nomination of the president of a chemical company that had been cited earlier as a polluter of the Houston Ship Channel. The appointment failed, primarily due to the efforts of citizens' groups concerned about the environment. But this powerful, unified political pressure from those affected by pollution was called forth only when such a clear choice was involved and public enthusiasm was aroused by the drama of a highly charged issue. On more routine matters, the board in its formative period weighed industry needs more heavily than those of less organized groups.[43]

Part of the problems involved in enforcing the new laws resulted from the weaknesses of the young agencies themselves. The air board's members sat without salary, and salaries were insufficient to attract highly qualified personnel for other staff positions. In general, these regulatory bodies had budgets that were far below the amount required to fulfill their stated purposes. In the crucial area of establishing enforceable norms, they lacked the expert personnel to establish and defend standards upon which the board members could agree.

Some of these difficulties were eased by the growing federal role in pollution control. Throughout the 1950s and 1960s, new laws gradually increased the national government's power in this area, and the implementation of these laws resulted in the expansion of a variety of pollution control agencies located in various bureaucracies throughout the federal government. This process of legal and institutional change climaxed in the late 1960s and early 1970s with a series of strong laws passed in the midst of widespread public concern for environmental quality. The lasting institutional legacy of this period was the Environmental Protection Agency, a powerful regulatory body created in 1970 by bringing together most existing federal pollution control agencies and then giving the single institution which resulted increased powers. The EPA quickly became the focus of pollution control initiatives. It put pressure on Texas agencies to establish and enforce stronger standards for allowable discharge while using its own resources to help determine what those standards should be. It made available financial and technical assistance to state agencies, contributing greatly to the growth of a "public" body of knowledge and expertise on the complex technical problems involved in determining how to reduce pollution. As a result of the efforts of the EPA and other federal agencies such as the Council on Environmental Quality, the public sector began to develop the capacity to deal critically and forcefully with the information and expertise made available to it by the more experienced, more specialized private sector.[44]

In addition to such federal leadership, pressures for improved state enforcement also came from the regional level, where pollution problems were still more easily identified. Harris County officials helped force the Texas Air Quality Board into action against flagrant violators by bringing suit against regional polluters. Jefferson County citizens' groups organized to protest the granting of air pollution variances to numerous local refineries and petrochemical plants. With the creation of regional air pollution control districts under federal authority, local initiatives became more systematic and more powerful. As public institutions at all levels gained experience, jurisdictional confusion decreased, and the public agencies could turn more of their resources to the problems of enforcing their regulations.[45]

Private organizations like the API and the large oil companies retained a strong interest in imposing their own standards on these public institutions, and the oil firms did not easily surrender their traditional autonomy in this area. In the 1970s, the emerging regulatory bodies sought to establish their power to enforce compliance, and a series of major disputes resulted. The Port Arthur refineries of Texaco and Gulf and Mobil's Beaumont plant were charged with violations of the clean air act. Responding to the pressure of federal deadlines, the state agency levied fines that were substantial in comparison to anything levied previously—if not in comparison to the companies' ability to pay. Unfavorable publicity seemed to bother the companies involved more that the fines, and they sought to counteract government charges by appealing to the public's desire for continued growth and for adequate supplies of gasoline. In their ongoing struggle over the definition and implementation of new standards, the oil companies' record of sustained growth in both quantity and quality of refined products was a politically potent and frequently cited argument.[46]

Despite this record, however, by the 1970s the balance between public and private control of the environment had begun to shift toward the public sector. The sustained growth of manufacturing and the increasing complexity and diversity of pollution had led to a series of crises that called forth demands for solutions. Reliance on private control of industrial pollution had proven inadequate; the resulting rise of public environmental planning bodies represented the most significant rearrangement during this century of the institutional framework within which responses to pollution took place.

Along with this shift came a new level of public interest in environmental problems. In 1922, water pollution in the Houston Ship Channel had been labeled a "trifling matter" by a prominent oil industry spokesman, and the large oil firms subsequently had contained most political efforts to impose controls. But in 1971, the head of Texaco's Port Arthur refinery voiced a far different attitude: "Our objective is to keep well within government air and water quality standards, and we will continue to correct any adverse conditions which may be attributable to our operations."[47] At about the same time, the president of Texaco cited the government's "major responsibility" of "setting standards based on realistic goals that have proved necessary to protect the nation's health and welfare."[48] Such statements were supported by substantial investments in antipollution equipment.[49] Pollution control was no longer primarily the private concern of those who produced refined goods. Rather, maturing national public institutions had taken control in establishing and enforcing standards of allowable discharge.

The tension between compliance with these standards and unrestrained expansion of productive capacities remained strong in this refining region, since any threat to the continued growth of the refineries was also a threat to the health of the entire regional economy and most of its population. The potential impact of strict pollution control on regional growth was evident in 1971, when the Texas Air Control Board shut down a small chemical plant in Port Arthur for failure to meet discharge standards. This order starkly contradicted the traditional commitment to growth regardless of costs. It was hardly a fair test of the depth of the public sector's resolve to control pollution, however, since the plant affected was a relatively small operation and a blatant polluter. Indeed, when Texaco and Gulf faced similarly stringent state rulings, local officials and even the governor of Texas supported their successful appeals for variances.[50]

Such compromises between the corporate autonomy of the not-so-distant past and demands for immediate solutions from the youthfully belligerent public agencies seem likely to characterize future pollution control efforts. Having recognized the government's authority to set and enforce standards, the private concerns nonetheless retain a substantial measure of influence, especially in establishing what level of pollution control is economically feasible.[51] For their part, public planning bodies continue to operate under several constraints inherited from an earlier period that will probably keep them from increasing to a significant degree their newly attained power. First, the tasks that they must perform with resources that are still very limited continue to mount. More importantly, environmental planning is a volatile political issue that is increasingly coming into conflict with other pressing political concerns, most notably the need for an adequate supply of energy and concern over sustained inflation. Finally, the environmental agencies appear to be vulnerable to a form of political emaciation. Support for their efforts could wane if they prove unable—even in the short run—to define and implement equitable programs of resource allocation and conservation.

No matter how well conceived, planning that is perceived by the polity as a threat to sustained growth faces severe difficulties. The public planning institutions became strong by responding to crises caused by the cumulative effects of a long lack of adequate controls on pollution. While easing the problems that they were created to deal with, however, these agencies may undermine their own position; once the short run crises are "solved," the society might well revert to its traditional stance of disregarding long run environmental problems. In that case, a paradoxical deadlock could result: several decades of experience will have prepared public institutions for environmental planning tasks in a political economy that is unwilling to utilize them.

Even if the long-standing devotion to unrestrained growth has been permanently tempered in the United States by the events of the last decade, a critical institutional barrier to pollution control remains. The growth of petroleum pollution has continued to mirror that of the oil industry. Local government could not respond to the new conditions created by the initial development of Spindletop. After several decades of growth, the state government no longer had the jurisdiction or power to deal with many of the problems associated with the giant refining complex and the expanding world petroleum industry of which it was an integrated part. Currently, the national government is powerless to control many of the international environmental consequences of the growing petroleum industry. The lag between public and private capacities remains; mounting pollution of the oceans and the atmosphere as increasing amounts of petroleum products are shipped over long distances could well lead to demands for stronger controls in a world without the institutional means to impose and enforce such controls.

NOTES

1. The creation of new institutions to deal with pollution was thus made difficult by the existence of institutions with a strong stake in controlling the extent of the response to pollution. In this respect, the situation was much like that faced by those who sought to organize independent industrial labor unions. In both cases, organizers had to attempt to fashion their own organizations from the ground up while challenging the power of large well-organized corporations with a strong incentive to maintain control.

2. The private institutions retained, of course, a substantial role in carrying out the

dictates of the public sector, but their power to determine the extent and the nature of responses to environmental problems waned with the maturation of public agencies.

3. Mayor Ben Campbell to Judge James Autry, letter dated August 12, 1913, in Box 30, Personal Correspondence (1913), papers of James L. Autry, Rice University, Houston, Texas. For a good summary of regional and state development prior to 1901, see John Spratt, *The Road to Spindletop* (Dallas, 1955).

4. James Clark and Michel Halbouty, *Spindletop* (New York, 1952), pp. 52–94, 105. Written descriptions of the waste that accompanied early production at Spindletop cannot match photographs. See Walter Rundell, Jr., *Early Texas Oil: A Photographic History, 1866–1936* (College Station, Texas, 1977).

5. The lack of reliable, systematic sources of information on the extent of pollution in this early period forces the historian to rely, perhaps excessively, on descriptive information—read in the context of recent advances in the understanding of pollution—in reconstructing conditions.

6. *Oil Investors' Journal,* July 5, 1902, p. 5; September 1, 1902, p. 8.

7. For the classic statement of the long-run costs of such attitudes, see K. William Kapp, *The Social Costs of Private Enterprise* (New York, 1950). See, also, Matthew Edel, *Economies and the Environment* (Englewood Cliffs, N.J., 1973) and Joseph Petulla, *American Environmental History* (San Francisco, 1977).

8. The response of many oil executives to pollution control in the 1970s is very similar to the response of earlier executives to the rise of the CIO in the 1930s.

9. Wallace Farnham, "The Weakened Spring of Government: A Study of Nineteenth-Century American History," in Stanley Katz and Stanley Kutler (eds.), *New Perspectives in the American Past* (Boston, 1972), pp. 28–44, provides insights into the inadequacies of late nineteenth century government. His general approach is also useful in understanding important shortcomings of the public sector in the twentieth century. See, also, Thomas Cochran, *Business in American Life: A History* (New York, 1972), chapters 12, 13, and 20.

10. Even these figures are deceptively high, since these funds were used for various purposes throughout the state, not just in the refining region. See *Annual Report of the Comptroller of Public Accounts of the State of Texas, Year Ending August 31, 1900* (Austin, 1901), pp. 107–124. For purposes of comparison, it is useful to note that about $40 million was invested by private companies in the Southeast Texas oil industry from 1901 to September 1904. See *Oil Investors' Journal,* September 1, 1904, pp. 1–3.

11. For a brief sketch of the histories of the national pollution and control and conservation agencies, see Cynthia Enloe, *The Politics of Pollution in a Comparative Perspective: Ecology and Power in Four Nations* (New York, 1975), pp. 178–189. For the Texas state agencies, see John T. Thompson, *Public Administration of Water Resources in Texas* (Austin, 1960), pp. 14–60.

12. For a pathbreaking treatment of conservation policy in the context of the evolving national political structure, see Samuel Hays, *Conservation and the Gospel of Efficiency* (New York, 1969). See, especially, the author's preface to the 1969 edition (Atheneum) of the book, which was originally published in 1959.

13. This confusion caused by inherent conflicts in jurisdiction could be exploited by large, nationally-active corporations which could attempt to locate the level of political power most open to their influence and then seek to have decisions reached at that level. The state of Texas, for example, was generally a more favorable political arena for the oil industry than was the national government.

14. Department of the Interior, U.S. Bureau of Mines, *Pollution of the Coast Waters of the United States* (Washington, 1923), p. 538. A copy is in the API library in Washington, D.C.

15. Ibid., pp. 1–10.

16. Ibid., pp. 434, 440, 64.
17. Ibid., pp. 435, 461, 462.
18. Ibid., pp. 577–578.
19. J. Clarence Davies III and Barbara Davies, *The Politics of Pollution,* 2nd ed. (Indianapolis, 1975), p. 38.
20. Benjamin Allin, *Reaching for the Sea* (Boston, 1956), pp. 80–103; B. C. Allin to Mayor of Houston, letter dated December 10, 1922, in "Houston Ship Channel" folder, Papers of Joseph Stephen Cullinan, University of Houston, Houston, Texas; W. W. Moore to Mayor Amerman, letter dated December 11, 1922, in "Houston Ship Channel" folder, Cullinan papers; memo to Joseph Cullinan, dated December 27, 1922, in "Houston Ship Channel" folder, Cullinan Papers. The *Houston Chronicle,* July 12, 1925, p. 8, records citizens' protests of continuing discharges of waste oil into the ship channel.
21. Bronson Batchelor, *Stream Pollution: A Study of Proposed Federal Legislation and Its Effect on the Oil Industry,* p. 51, manuscript dated February 1937, in the API library, Washington, D.C.; and Bureau of Mines, *Pollution of the Coast Waters of the United States,* pp. 1–44.
22. W. B. Hart, "Controlled Disposal of Wastes Versus Pollution," *Oil and Gas Journal,* May 14, 1936, p. 234. For a general discussion of these bills, see Herman Baity, "Aspects of Governmental Policy on Stream Pollution Abatement," *American Journal of Public Health* (December 1939), pp. 1297–1307.
23. Bronson Batchelor, *Stream Pollution,* pp. 45–51.
24. Ibid., p. 34. A broad study of the oil industry's experience with such "political factors" is Robert Engler, *The Politics of Oil* (Chicago, 1961). For a general review of national oil policy, see Gerald Nash, *United States Oil Policy, 1890–1964* (Pittsburgh, 1968). For a general review of air and water pollution legislation, see Clarence Davies III and Barbara Davies, *The Politics of Pollution.* Also, see Frank Smith, *The Politics of Conservation* (New York, 1966).
25. *Beaumont Enterprise,* March 26, 1939, p. 1; November 11, 1939, p. 4; January 24, 1940, p. 9; January 26, 1940, p. 16.
26. W. B. Hart, "Controlled Disposal of Wastes Versus Pollution," p. 237.
27. API, "Report Covering Survey of Oil Conditions in the United States," 1927, p. 1. Report on file in API library, Washington, D.C.; API, *Proceedings of the 17th Mid-Year Meeting, Division of Refining,* San Francisco, May 12–15, 1955, pp. 299–300.
28. W. B. Hart, "Disposal of Refinery Waster Waters," *Industrial and Engineering Chemistry* 26, no. 9 (September 1934), p. 965.
29. API, *Manual of the Disposal of Refinery Wastes, Section I: Waste Water Containing Oil* (New York, 1941); *Section II: Waste Gases and Vapors* (New York, 1931); *Section III: Waste Waters Containing Solutes* (New York, 1935).
30. W. B. Hart, "The Waste Control Laboratory of the Atlantic Refining Company," *Water Works and Sewerage* 88 (January 1941): pp. 30–31.
31. For a brief discussion of tetraethyl lead, see *Oil and Gas Journal—Oil City Derrick,* Diamond Jubilee Issue, August 27, 1934, p. 199. For a general discussion of the cumulative effects of such lags, see Barry Commoner, *The Closing Circle: Nature, Man and Technology* (New York, 1971). See, also, W. B. Hart, *Industrial Waste Disposal for Petroleum Refineries and Allied Plants* (Cleveland, 1947); W. B. Hart, "Disposal of Refinery Waste Waters," p. 967; C. G. Rebman, "Elimination of Waste in Refining," *Oil and Gas Journal,* November 4, 1926, pp. 126–130. The general advances in refining technology are discussed in John Enos, *Petroleum Progress and Profits: A History of Process Innovation* (Cambridge, Mass., 1962).
32. *Oil and Gas Journal,* November 4, 1926, p. 126; *The Look Box* 3, no. 8 (August 1921), p. 6; *Oil and Gas Journal,* January 13, 1922, pp. 25 and 27. See, also, James H. Westcott, *Oil,*

Its Conservation and Waste (New York, 1928); Clifford Russell, *Residuals Management in Industry: A Case Study of Petroleum Refining* (Baltimore, 1973).

33. Bureau of Mines, *Pollution of the Coast Waters,* pp. 42–43. Of course, "noneconomic" factors—a court case, political lobbying, public relations—could represent substantial "economic" costs to the firm.

34. See "Port Arthur Scrapbooks," vol. II, index number 18, at Gates Memorial Library, Port Arthur, Texas; C. Rebman, "Elimination of Waste in Refining," p. 130; Texaco, *Protecting the Environment,* 1973, p. 7.

35. W. B. Hart and B. F. Weston, "The Water Pollution Abatement Problems of the Petroleum Industry," *Water Works and Sewerage* 88 (May 1941), p. 217, suggest that improvement occurred throughout the nation in waters affected by the oil industry in the decade before 1941. For a general survey of national conditions in the 1930s, see U.S. Natural Resources Committee, Special Advisory Committee on Water Pollution, *Report on Water Pollution* (Washington, 1935).

36. Texas Highway Department statistics record an increase of more than 500 percent in automobile registrations in Harris County (Houston) from 1947 to 1970.

37. Southwest Research Institute, "Project 566-1: Air Pollution Survey of the Houston Area," dated July 1, 1958, pp. ii–x, on file at Houston Chamber of Commerce, Houston, Texas; *Beaumont Enterprise,* July 25, 1952, p. 10; September 6, 1962, p. 9; September 26, 1952, p. 1; September 22, 1955, p. 11.

38. Edward B. Williams, Jr., "Pollution Control: A Houston Ship Channel Issue," master's thesis, Political Science Department, Texas A&M University, August 1972, pp. 34–47.

39. Ibid., p. 48.

40. Southwest Research Institute, "Project 21-1587: Air Pollution Survey of the Houston Area, 1964–1966," dated October 1966, p. vi, on file at Houston Chamber of Commerce, Houston, Texas.

41. *Houston Post,* September 13, 1967, p. 1.

42. G. Todd Norvell and Alexander Bell, "Air Pollution Control in Texas," in *Legal Control of the Environment* (New York, 1970), pp. 337–376.

43. Ibid, p. 355; Edward Williams, "Pollution Control," pp. 71–76, argues that "the official composition and actual membership on the Board made its position inevitably pro-industry" (p. 71).

44. For a general overview of the growth of the federal government's powers to control pollution, see J. Clarence Davies III and Barbara Davies, *The Politics of Pollution.* For a fascinating insider's view of the early years of the EPA, see John Quarles, *Cleaning Up America* (Boston, 1976). G. Todd Norvell and Alexander Bell, "Air Pollution Control in Texas," p. 373. For a detailed account of another state's difficulties in passing and enforcing environmental protection legislation, see Marc Landy, *The Politics of Environmental Reform: Controlling Kentucky Strip Mining* (Baltimore, 1976).

45. G. Todd Norvell and Alexander Bell, "Air Pollution Control in Texas, " pp. 368–373; *Beaumont Enterprise,* November 5, 1969, p. 1. The Harris County Air Pollution Control District was created in 1967; a Beaumont-Port Arthur-Orange District in 1969.

46. For such offenses, the possible fines were $1,000 per day per offense. In 1972, Texaco's Port Arthur refinery was fined $6,000 by the Texas Air Control Board. Great Lakes Carbon in Port Arthur paid a $25,000 fine for more severe violations in the same year. See *Beaumont Journal,* July 4, 1972, p. 1; April 4, 1972, p. 1; *Beaumont Enterprise,* July 27, 1973, p. 1.

47. Press release to the *Port Arthur News,* CavOilcade edition, 1971, p. 3. Xerox copy from the Refining Department of the Texaco Archives.

48. Texaco, *Protecting the Environment,* p. 3.

49. API estimates of oil industry environmental expenditures suggest an increase of nearly 500 percent between 1966 and 1974, with the total of such spending in the nine-year period surpassing $7 billion. See API, *Environmental Expenditures of the United States Petroleum Industry, 1966–1974* (Washington, 1975). See, also, Council on Economic Priorities, *Cleaning Up: The Cost of Refinery Pollution Control* (New York, 1975).

50. *Beaumont Enterprise,* July 29, 1971, p. 2A; Gregg Kerlin and Daniel Rabovsky (Council on Economic Priorities), *Cracking Down: Oil Refining and Pollution Control* (New York, 1975), pp. 394–395.

51. For example, see Earl Oliver, Robert Muller, and F. Alan Ferguson, *The Economic Impact of Environmental Regulations on the Petroleum Industry* (Menlo Park, Ca., 1974). This is a report prepared for the API by the Stanford Research Institute.

Conclusion

Two characteristics of twentieth century development on the upper Texas Gulf Coast have been the focus of this history: the sustained oil-led industrialization of the region and the evolution of an institutional nexus of growth composed of large-scale, oil-related organizations. In the overview of industrialization presented in Part I, a variety of statistics ranging from total exports to the number of industrial workers employed suggested the regional importance of the oil industry in general and of petroleum refining in particular. No such statistical summaries were available with which to sketch the outlines of institutional change in Part II. Instead, I have sought to describe the changing institutional framework that shaped the development of the region's oil industry by recounting case histories of many of its most important parts. These case histories have not added up to a comprehensive "institutional" history of the region, but they have offered historical perspectives on the changing roles of the major oil companies, the refinery unions, and the various government agencies that both promoted and regulated the oil industry. Taken as a whole, these case studies provide a rough, incomplete sketch of the institutional changes that accompanied twentieth century development in the refining region.

The central institutional development in the years immediately after the discovery of the Spindletop oil field was the creation of large, vertically integrated oil companies that built their primary refineries in the region. In their formative years, these companies faced severe economic and political challenges. The first and most basic was simple: they had to build organizations capable of surviving in the highly competitive and chaotic Texas oil fields while also taking steps to insulate themselves from the ever-present threat of Standard Oil, which then dominated the national oil industry. In seeking economic advantages, Gulf Oil and Texaco quickly sought to replicate the vertically integrated organizational structure of Standard Oil. Although this helped them to meet economic challenges, it also opened them to political attacks. Antitrust laws and incorporation laws written before the discovery of oil in Texas combined with an almost hysterical fear of Standard Oil to focus political attention on any company thought to have ties with Standard Oil. As the largest regional companies, Texaco and Gulf faced especially strict scrutiny. Oil-producing companies often led this outcry against their larger, vertically integrated rivals in an effort to secure their own competitive positions. Thus in the early years of regional oil development, the major Gulf Coast oil companies had to overcome laws and attitudes inherited from the pre-oil period while seeking to meet competitive threats posed by other regional companies and build vertically integrated organizations capable of competing with Standard Oil in national and international markets.

Yet in institutional terms, this early era was not simply one of challenges to the emerging oil companies, for they came to enjoy a great deal of autonomy in many aspects of their operations. They exercised direct or indirect control over a variety of oil-related supply industries, since such close ties were often essential in assuring adequate supplies of the products needed to build and operate their production, transportation, and refining facilities. More lasting than the oil companies' control over secondary supply industries was their almost absolute control of their labor forces. No strong or independent workers' organization existed to push for improvements in the lot of the individual worker or to present demands in the collective interest of the entire work force. On at least one occasion when groups of workers sought to organize, the state government promoted the continued dominance of management over labor by providing "public" force on the side of the companies in their disputes with labor. Such instances were uncommon, however, since the companies generally quieted discontent with wages that were higher than those paid in other regional industries. The government officials and agencies that controlled river and harbor improvements were apparently also susceptible to "high wages" of another sort, and large campaign contributions, intensive lobbying, and perhaps even outright bribery helped regional oil interests to secure the improvement of waterways required for the operation of the large tankers used in shipping petroleum products. Such transportation improvements were, of course, often useful to the region as a whole, but the economic importance of the major oil companies made them especially effective in obtaining such promotional services from the government.

In this early, transitional period in the history of the refining region, the emerging regional oil companies ultimately proved effective in controlling most oil-related matters. After withstanding antitrust attacks and rebuking the competitive assaults of other oil companies, the largest firms that survived gradually asserted tighter control over both their economic and political environments. Indeed, as World War I began, these companies had grown into internationally active, vertically integrated concerns that wielded substantial power in the region from which they had grown. As the dominant modern, large-scale economic institutions active in the region, these oil companies controlled much of the economic life of the upper Texas Gulf Coast. Due to the lack of labor unions, they also controlled the conditions of labor in the most substantial industrial concerns in the region, the refineries. Finally, public sector agencies as yet had neither the clear-cut legal authority nor the institutional capacity to regulate the operations of the large oil companies. All in all, the major oil companies in these early years were in an enviable situation: they exercised close control over almost everything that was of importance in their operations.

This began to change gradually in the middle years (1916–1941). The most striking institutional change in this period was the emergence of independent, industrial labor unions in the coastal refineries. Building on precedents set during World War I, refinery workers began to attain better working conditions and even a semblance of self-representation in the 1920s. Although the company unions of that decade did not generally survive the Great Depression, they laid the foundation for the independent labor unions that were organized in the tense and often violent 1930s. On this foundation, the CIO, with the aid of the National Labor Relations Board, built independent unions for the region's largest group of industrial workers, and the creation of such organizations was an institutional innovation of considerable importance in the region's history. Where the earlier company unions had been extensions of the corporations that established them, the AFL and CIO unions were tied to national workers' organizations that were independent of the oil companies. This made a crucial difference in the lot of the worker, who came to enjoy greater job security and job control in addition to wages and benefits that remained high relative to other regional industries.

The emergence of independent unions in the middle period was part of a general trend toward the gradual weakening of the major oil companies' direct control over aspects of oil-related development. As the Gulf Coast became more completely integrated into the national petroleum industry, the oil companies that had been born at Spindletop could increasingly rely on independent supply firms and nationally active construction specialists for many of the goods and services that they had been forced by circumstances to supply for themselves in the earlier period. This differentiation of tasks within the oil industry meant that the large oil companies could concentrate almost exclusively on expanding their vertically integrated operations while relying on other concerns for crucial secondary functions, and this resulted in a growth in the number of oil-related concerns in the regional economy that were not directly controlled by the major companies.

In this same period, the oil-related segments of the public sector increased in importance, performing certain crucial functions that the companies alone could not fulfill. River and harbor improvements continued; the opening of the Houston Ship Channel, the building of an intercoastal canal, and the further deepening of the region's harbors all were of special importance to the region's oil industry. In the 1920s and 1930s, however, several public agencies came to play new and equally important roles. The first such agency was the U.S. State Department. As the Gulf Coast refineries began to use Mexican and Venezuelan oils, the complex relationships between the oil companies and these foreign governments became an important determinant of regional development, and the State Department took a central part in these on-going negotiations. In the early negotiations with Mexico, the State Department behaved almost as an adjunct to the oil companies, but it gradually grew more independent as it gained in experience and as it matured as an institution. A second important "promotional" agency that grew in importance in these middle years was the Texas Railroad Commission, which took a leading role in the stabilization of crude oil supply—and crude oil prices. As the manufacturing branch of vertically integrated companies active in all aspects of the petroleum industry, the coastal refineries were affected by institutions and events that influenced all parts of the operations of their parent companies. Government efforts to help these companies secure adequate and stable supplies of crude oil were crucial to the health and continued growth of the refining region, since the steady expansion of the giant coastal refineries required ever-greater supplies of crude oil.

The public sector's involvement in the oil industry in the middle period was more promotional than regulatory. After World War I, the vigorous antitrust crusades of the prewar years subsided, and subsequent efforts at regulating aspects of the oil industry's performance proved halfhearted. When oil pollution of water became a serious problem in the 1920s, for example, the public sector's response was a weak law that did little more than acknowledge the problem. Not until the Great Depression radically altered the political climate of the 1920s did government regulatory efforts have a significant impact on the refineries on the Gulf Coast. In the 1930s the National Labor Relations Board intervened forcefully and decisively in labor disputes in the region's refineries, paving the way for the victories of the CIO and undermining the already tenuous position of existing company unions. Two firsts in the institutional history of the region's oil industry thus occurred simultaneously in the late 1930s: the successful intervention of an independent government regulatory agency and the creation of a strong independent worker's organization. Both changes illustrated trends that were to become more apparent after the war, the growing power of national institutions in regional affairs and the emergence of new institutions capable of challenging the authority of the oil companies on a variety of oil-related matters.

Both of these trends continued in the modern era, contributing to the economic and institutional diversity of the region. The growth of new economic activities, notably the production of petrochemicals and of primary metals, generated whole new industrial complexes. Each required its own secondary supply industries while giving rise to new labor organizations and encouraging new promotional and regulatory roles for government agencies. This increasing diversity was reenforced by on-going changes within the petroleum industry. In the modern era the independent labor unions, which had gained footholds in the coastal refineries in the 1930s, matured into economically and politically powerful organizations capable of challenging management on those issues that most directly affected their members. The large oil companies also came to share control over other aspects of oil-related development, as new government agencies began to regulate hiring practices, environmental problems, and, finally, the prices paid for crude oil. The vertically integrated companies that owned the large coastal refineries remained central to the region's development, but in the years after World War II, they no longer enjoyed the nearly total control over oil-related matters that they had exercised in earlier years. Although far from powerless, these companies nonetheless had to learn to share authority with other independent institutions in an increasingly complex political and social environment.

To understand these new institutions, however, it is important to bear in mind certain limits to their independence. Many of them retained close ties to the oil companies. This was clearest in the numerous close linkages between petroleum and the largest new industry in the region, petrochemical production. Labor organizations had a somewhat similar identity of interest with refinery management on any issue that affected the growth of the industry and therefore the size of the work force. In addition, the position of the unions was being weakened by the impact of technological change. In the political realm, regulatory agencies remained partially dependent on the companies for expertise and information; they were also susceptible to the indirect influence that arose from the oil companies' involvement in electoral politics. Finally, the actions of all of these organizations were constrained by a broadly shared societal belief that the center firms in oil, whatever their faults, were the primary sources of regional growth—and growth remained throughout the century the most widely shared economic goal of the society.

A final constraint was historical: because the large oil companies fashioned the first powerful, modern organizations active in the region's oil industry, they could exert considerable influence on the pace and timing of the emergence of subsequent oil-related institutions. Although those who ran the oil companies repeatedly sounded the rhetoric of laissez faire, their frequent use of the promotional services of public agencies hastened the expansion of those agencies. Secondary economic institutions—from supply firms to construction companies—also grew rapidly, as the largest oil companies generally encouraged their growth in order to focus their energies on the primary branches of the petroleum industry. Perhaps most severly hampered by the power of the oil companies was the rise of independent labor unions. Faced with the often vehement and even violent opposition of management, these organizations only gradually became able to assert their independence. Such was also true of government regulatory agencies. Efforts to enforce the antitrust laws in the first decade of the century, the activities of the National Labor Relations Board in the 1930s, and the more recent attempts to formulate and implement national environmental and energy regulations all met the aggressive opposition of the oil companies, which generally viewed such regulatory efforts as unwarranted challenges to their traditional autonomy. Such opposition did not prevent the creation of either independent regulatory commissions or independent labor unions, but it did slow the emergence of these institutions.

This was especially clear in the neglect of oil-related pollution. Since at least the 1920s, groups outside the industry had sought to convince the public sector that such problems required politically imposed solutions.

But for most of the period, the large oil companies defined the role of the public sector, as well as the private, in matters relating to oil pollution. Their large plants operated more efficiently and were therefore cleaner than those of smaller manufacturers, but the scale of their operations also made them the primary sources of pollution. After this contamination presented a problem so serious that it could no longer be ignored, the largest oil companies still retained a measure of control over the political institutions that emerged to formulate and implement environmental policy. A historical view of the responses to petroleum-related pollution in the coastal region suggests several paradoxical conclusions. For sixty years the refineries polluted with impunity and reaped the benefits of a production process that was used without taking into account the total societal costs of refining. After enduring oil-related pollution thoroughout this period, the remainder of the society then helped pay with its tax dollars for the necessary controls and for the correction of past abuses. The center firms in oil, which had profited in the past from their neglect of the environmental implications of their unrestricted production, understandably objected to this "unwarranted" intrusion of government into the affairs of private enterprise; the private solutions of the past, after all, had worked very well from their perspective, though not from that of the larger society.

Public responses to pollution and to other problems related to the growth of refining were hampered by one of the basic differences between the political institutions and the oil companies: the expansion of the center firms in oil was not bound by the type of jurisdictional boundaries that characterized the political system. Public responses to the growth of these businesses could not therefore take the same scale as the problems involved. The large corporations planned their activities on an international scale, but political agencies were, at most, national, and even within the nation there were smaller divisions in authority that hampered the formulation of systematic policy.

As the public sector sought to overcome such problems, the growing complexity of the political economy of oil gradually blurred the once sharper distinction between public and private institutions. The Texas Railroad Commission, for example, performed an essentially economic function, and its use of its power over the level of production after the 1930s made it a "semi-private" institution. On the other hand, the American Petroleum Institute's close cooperation with several government bureaucracies and its role as political lobbyist for the petroleum industry meant that it often seemed to be "semi-public." The simpler, traditional distinction between public and private functions failed to capture the growing diversity of the increasingly complex institutional arrangements of the modern era.

This diversity made the phrase "refining region" less than complete in its characterization of the upper Texas Gulf Coast. The refining region had been created originally in the two decades after the Spindletop discovery, as giant refineries spread along the coast in the Port Arthur area and then along the banks of the Houston Ship Channel. These plants gave the two areas a common industrial base which expanded rapidly in the years between the two World Wars. In this heyday of regional refining, the petroleum refineries were the largest industrial concerns on the upper Texas Gulf Coast, and the vertically integrated companies that owned them were the most important economic institutions active in the region. The maturation of the regional economy during and after World War II, however, reduced both the economic role of refining and the power of the largest oil companies relative to that of other institutions. In the last half of the twentieth century, the region became less distinct from the remainder of the booming coastal plain of Texas and Louisiana. At the same time, refining lost its earlier position of dominance, becoming simply one major industry in an increasingly diverse economy.

Its historical imprint nonetheless remained. The images that have marked the various phases of regional development in the national consciousness embodied the continuing significance of petroleum to the region. The Spindletop gusher, which first attracted the attention of the nation to the Texas Gulf Coast, became the symbol of the first decades of regional development. Then as refineries and petrochemical plants crowded the coast from Port Arthur to the Houston area in subsequent years, they became the new symbol of the industrialization of the region. In recent years, even the tall towers of the refineries have been overshadowed by the sprawling skyline of Houston, and the skyscrapers that house the administrative headquarters of numerous petroleum, natural gas, and petrochemical companies have become the new national symbol for the expansive economy of the upper Texas Gulf Coast. Behind each of these symbols lies the historical reality of almost eighty years of oil-led growth that has transformed a gradually industrializing timber and cotton producing area into a major urban-industrial region closely tied into the national economy. Whatever their evaluation of the consequences of this on-going transformation, those who have lived through it and those who have studied it must shake their heads in amazement at its rapidity and thoroughness.

Appendix I
Gulf Coast Refining Capacity Before 1918

Since 1918 the United States Bureau of Mines has published systematic statistical information for each individual refinery in the nation. Although the yearly statistics have probably omitted some of the smaller refineries, especially those that were in operation for less than a year, they provide a dependable series of information concerning petroleum refining. The historian is not so fortunate, however, when he ventures back beyond 1918. In attempting to reconstruct the capacities of refineries on the Gulf Coast between 1901 and 1918, I have relied on a variety of sources, some more dependable and more complete than others. In this appendix, I will review the information that I have used in order to arrive at approximations of the capacities of each of the major Gulf Coast refineries before 1918.

The totals for the Gulf refinery at Port Arthur were the most difficult to assemble. John McLean and Robert Haigh, *The Growth of Integrated Oil Companies* (Boston, 1954), p. 682, gives information about the integrated activities of Gulf throughout its history. The listing of "Domestic Refinery Runs," the source of which is listed as "Gulf Oil Corporation," is equivalent to a listing for the Gulf Port Arthur refinery until 1911, when Gulf opened a second, smaller refinery in Fort Worth. Additional information about Gulf comes from three separate sources that were useful for all regional refineries. In 1904–1905, the Bureau of Corporations made a thorough investigation of the oil industry. The records that were used to compile the published volumes, *Report of the Commissioner of Corporations,* 2 vols. (Washington, 1907), are in the National Archives in

Record Group 122, the records of the Federal Trade Commission. These records include detailed statistical information by company and interviews with various refinery managers. They seem to be comprehensive, since numerous reports on small refiners are included. A later FTC report, known as the Oklahoma Oil Investigation, is less comprehensive. The background material for this report is also in the National Archives. It covers the years 1909-1914, although there is systematic information for only Gulf and Texaco. Included are useful statistics concerning costs, sources of crude, yield of products, and various other information. The final source for Gulf capacity was the *Oil Investors' Journal,* whose articles and advertisements provided bits of useful information.

From these varied sources, the following totals for the Gulf Oil Port Arthur refinery are taken: (in 42-gallon barrels per day):

1902—3773	McLean and Haigh (hereafter, M-H)
1903—6323	M-H
1904—7970	M-H, Bureau of Corporations (hereafter, BC), Drawer 352, File 2881, Part 1.
1905—10748	M-H, BC
1906—10767	M-H
1907—10277	*
1908—13019	*
1909—19540	FTC, File 1434-1, Schedule of Inquiries—Gulf Refining Company. Also M-H.
1910—24932	FTC, M-H
1911—27397	FTC, M-H
1912—26575	FTC, M-H

After 1913, the figures from McLean and Haigh include the output for Gulf's Fort Worth refinery and are not an accurate measure of the Port Arthur plant's output. All of the above figures represent crude actually consumed, which is usually less than the refineries' rated capacity. The capacity of the Gulf refinery is suggested by a Bureau of Corporation's estimate of 10,000 b/d in 1905, an FTC estimate of 33,000 b/d in 1913, and Gulf Oil advertisements in the *Oil and Gas Journal* in 1911 (30,000 b/d) and 1916 (55,000 b/d).

The totals for the Texaco refineries at Port Arthur and Port Neches are much more systematic. They are taken from information contained in the Texas Company Archives in White Plains, New York. A folder entitled "The Port Arthur Works" contains yearly capacities and outputs. Testimony submitted to the Temporary National Economic Committee in 1938-40 contains additional information. The crude consumption and capacity of each refinery were as follows (in barrels per day):

	Port Arthur		*Port Neches*	
	Capacity	*Runs to Still*	*Capacity*	*Runs to Still*
1903	1000 b/d	118 b/d		
1904	1000	872		
1905	2000	1502		
1906	2000	1762	3000 b/d	773 b/d
1907	2000	1643	3000	2219
1908	2000	987	3000	2367
1909	3000	2581	3000	2761
1910	5000	3882	3000	2657
1911	5000	3355	3000	2797
1912	7500	6773	3000	1503
1913	15000	11824	3000	2253
1914	20000	16603	3000	2689
1915	32000	25776	3000	3170
1916	32000	26019	8000	7017
1917	32000	29814	13000	11173
1918	32000	28595	13000	10649

The Bureau of Corporations and FTC investigations contain additional information on such things as costs and yields. But for all of the aspects of Texaco that I studied, the best source was the T.N.E.C. submittal.

No such systematic information was available on the operations of the Security refinery at Beaumont (later Magnolia and finally Mobil). The early years of this refinery were clouded by its problems with the Texas antitrust laws, and information on operations, like operations themselves, was sporadic. The Bureau of Corporations reports a daily crude consumption of 4392 in 1904 (see Drawer 352, File 2881, Part 2), and a capacity of about 4000 b/d for the year 1905. Information from then until 1912 is very sketchy. In 1909 the *Oil Investors' Journal* reported a capacity of 8000 b/d for the refinery (see *OIJ*, August 6, 1909, p. 27). Additional information about the years from 1903–1907 comes from the *United States* v. *Standard Oil Company of New Jersey, et al.* (U.S. Circuit Court, Eastern Division of Eastern Judicial District of Missouri, 1906) 21 volumes. The yearly output of the refinery is given in Petitioner's Exhibit #369, volume 8, p. 897. After the formation of the Magnolia Refining Company in 1911, information becomes more complete. *The History of the Magnolia Refining Company* (Beaumont, Texas, n.d.) lists the plant's crude runs for 1912 at 7984 b/d and 1914 at 12,104 b/d. This account also suggests a capacity of about 25,000 b/d in 1916. The final source of use in reconstructing the capacity of the refinery is the FTC investigation. Files 1434-3 and 7336-185 contain detailed information on all phases of the plant's operation and ownership. FTC records list a capacity of 32,000 b/d for Magnolia's Beaumont works and a crude consumption of 7696 in 1913 and 28832 in 1918.

The last major refinery in operation in the region before 1918 was the Standard of New Jersey-affiliated refinery at Baton Rouge. Pre-1918 totals for this plant are available from the four volume *History of Standard Oil Company (New Jersey).* The following totals come from Hidy and Hidy, p. 414 and Gibb and Knowlton, p. 678:

1911— 7139	1915—19000
1912—11600	1916—23600
1913—19800	1917—35900
1914—17900	1918—43800

All of the above are estimated crude capacities in barrels per day.

Once the capacities of the region's five largest refineries have been estimated, the researcher enters a nebulous second level of information concerning the operations of the region's smaller refineries in the years before 1918. The historical record in this case is incomplete. Perhaps the two best sources for information are the Bureau of Corporations Petroleum Investigation and the FTC investigation, but even these are by no means comprehensive. Recognizing the limits of my data base, I make no claims that I have reconstructed a description of the entire Gulf Coast refining complex. In fact, I have attempted instead to include the refineries that were immediately below the largest five in terms of capacity and then to estimate the probably capacity of those numerous small refineries operating on the Gulf Coast.

The Bureau of Corporations' investigation includes information on United Refining Co. at Beaumont (capacity about 1000 b/d in 1903 with crude runs of about 110 b/d in 1904, see Drawer 451, File 3390); Southwestern Oil Company in Houston (capacity of about 300 b/d in 1902 and crude runs of 120 b/d in 1904, see File 3390); Record Oil Company in New Orleans (110 b/d crude run in 1904, see File 3279); and New Orleans Refinery and Petroleum Company (capacity listed as 10,000 b/d in 1904, but only about 14 b/d run to still). The *Oil Investors' Journal* contains fragmentary information concerning numerous small refineries with limited life spans during the years immediately after the Spindletop gusher. It is clear that although many companies made elaborate plans to refine Spindletop oil during the boom years, few actually built refineries that operated for any significant time.

The Record refinery apparently grew considerably between 1904 and 1910. FTC materials in File 7336-17-1 give a capacity of 10,500 b/d for the refinery, but there is no record of its actual operation. The FTC records also contain information concerning several refineries constructed in the years from 1914–1918. The Mexican Petroleum Company built a large plant at Destrehan near New Orleans in 1916 and 1917. Its capacity in December, 1917, was 800 b/d (see File 7336-34). The Freeport and Tampico Fuel Oil Company built a refinery at Meraux, also near New Orleans, in 1915 and 1916. Its capacity was listed as 10,000 b/d by the FTC in 1917 (File 7336-96). The FTC also gives information on Liberty Oil Company's New Orleans refinery, which opened in 1915 with a capacity of 700 b/d (File 7336-68-1) and on the Petroleum Refining Company's 1000 b/d refinery (as of 1917) near Houston (File 7336-257).

There is evidence that this accounting does not include several regional refineries. In 1916, H. G. James published a list of refineries throughout the nation, with information about their capacities as well as the year in which they were built. His list for Texas and Louisiana includes several Gulf Coast refineries not included in the FTC Oklahoma Oil Investigation records. He lists a Pure Oil Refining Company plant of 200 b/d capacity constructed near Houston in 1916, a 3000 b/d Pierce-Fordyce refinery built at Texas City in 1911, and a Pelican Oil Refining Company refinery at Chalmette, near New Orleans, built in 1915 with a 300 b/d capacity. Also included on James' lists is a large (10,000 b/d) Corona Oil Company refinery being constructed at New Orleans in 1916. James' figures do not correspond to those of the FTC in several instances, and they should be treated with skepticism. But they do indicate the existence of several refineries not reported in other sources (see H. G. James, *Refining Industry of the United States* [Oil City, Penn., 1916], pp. 6–11).

Pre-1918 refining statistics in general should be treated with a like measure of skepticism. From the variety of sources cited above, I have attempted to reconstruct the magnitude of refining activity on the Gulf Coast before 1918. By rounding my calculations to the nearest thousand barrels per day, I have indicated that they should be treated as very rough estimates. But even the yearly Bureau of Mines statistics after 1918 should be considered as good estimates, not absolutely exact figures. This is probably as accurate as anyone attempting to calculate over time the combined outputs of numerous refineries, many of which are highly unstable as to ownership and operation, can hope to be. My figures for the period before 1918 are, I think, good enough estimates to provide useful measures of the magnitude of refining activity and of the changes that took place over time.

Appendix II Classification of Refinery Owners by Size

LARGEST

	Company	*Refinery Location*	*Years in Operation*	*Greatest Capacity (Through 1961)*
1.	Gulf Refining Co.	Port Arthur	1902-present	269,000 b/d
2.	Texaco	Port Arthur	1903-present	280,000
		Port Neches	1906-present	40,000
3.	Mobil (Magnolia)	Beaumont	1903-present	200,000
		Magpetco	1924-1938	10,000
4.	Exxon (Standard-N.J.)	Baton Rouge	1911-present	356,000
		Baytown (Humble)	1921-present	292,000
		Ingleside	1928-1946	30,000
5.	Sinclair	Corpus Christi	1947-present	29,500
		Houston	1920-present	143,000
		Meraux, La.	1921-1934	22,000
6.	Shell	Deer Creek	1929-present	128,000
		Norco, La.	1930-present	102,000
7.	Standard of Indiana (Pan American)	Texas City	1934-present	103,000
		Destrehan, La.	1935-1959	33,000

I have used the term "Largest" to designate the regional refineries owned by the largest eight oil companies (as of 1946). "Years in operation" actually represents the years in which each refinery was listed in the Bureau of Mines Information Circulars, at least after 1918.

As is evident from the above chart, these refineries have enjoyed a record of stable operations. The four listed which are not shown as in operation as of 1961 were all the secondary regional refineries of one of the companies.

LARGE

	Refining Company	*Refinery Location*	*Years in Operation*	*Greatest Capacity on Gulf Coast*
1.	Pierce-Fordyce Co.	Texas City	1913–1929	10,000 b/d
2.	Mexican Petroleum Co.	Destrehan, La.	1916–1934	30,000
3.	Freeport and Mexican	Meraux, La.	1915–1920	10,000
4.	Galena-Signal	Galena	1920–1928	12,000
5.	Island Rfg. Co.	Sarpy, La.	1921–1926	10,000
6.	New Orleans Rfg. Co.	Norco, La.	1921–1929	20,000
7.	Pure Oil Co.	Nederland	1924–present	80,000
8.	Crown-Central	Pasadena	1924–present	40,000
9.	Atlantic Rfg. Co.	Port Arthur	1937–present	62,000
10.	Taylor Rfg. Co.	Corpus Christi	1939–present	45,000
		Port Isabel	1954–present	11,500
11.	Pontiac Rfg. Co.	Corpus Christi	1938–present	50,000
12.	Southwestern Oil & Rfg.	Corpus Christi	1937–present	50,000
13.	Eastern States Pet. Co.	Houston	1937–present	60,000
14.	Republic Oil Rfg. Co.	Texas City	1932–present	45,000
15.	Continental Oil Co.	Lake Charles, La.	1941–present	53,000
16.	Cities Service	Lake Charles	1942–present	185,000
17.	Phillips Oil Co.	Sweeney	1948–present	95,000
18.	Bay Petroleum Co.	Chalmette, La.	1951–present	40,000
19.	Suntide Rfg. Co.	Corpus Christi	1953–present	55,000
20.	Texas City Rfg. Co.	Texas City	1948–present	35,000

"Large" refineries are those that grew to the second level of capacity, immediately below that of the Largest. The Cities Service Lake Charles refinery (#16) is by far the largest in this group. However, it did not grow to its present scale until the 1950s; and for the historical overview of this chapter, I have therefore treated it as a member of the Large group.

All of the refineries listed above, except for those that grew in response to the availability of Mexican crude, have exhibited sustained growth.

SMALL

Houston Area:

Refining Company	*Refinery Location*	*Years in Operation*	*Greatest Capacity on Gulf Coast*
Hoffman Oil & Rfg. Co.	Houston	n/a–1920	1,000 b/d
Petroleum Rfg. Co.	Houston	n/a–1919	1,000
Freeport Gas Co.	Bryanmound	1920–1930	9,000
Terminal Oil & Rfg. Co.	Texas City	1924–1930	10,000
LaPorte Oil & Rfg. Co.	Morgan's Point	1920–1924	1,000
White Oil Corp.	Houston	1920–1922	5,000
Transatlantic Pet. Co.	Houston	1920–1924	1,000
Deepwater Oil	Houston	1921–1933	3,000
Keen & Woolf	Clinton	1921–1928	1,500
Penn-Texas Rfg. Co.	Morgan's Pt.	1921–1924	500
Turnbow Oil	Houston	1921–1922	1,200
Port Houston Oil & Rfg.	Pasadena	1921–1922	200
Mogul Producing & Rfg.	San Jacinto	1921–1922	600
Able Rfg. Co.	Houston	1922–1922	500
Cox Realization Co.	Humble	1922–1922	800
Deepwater Oil Refineries	Morgan's Pt.	1924–1927	800
Houston Terminal Rfg.	Houston	1926–1929	3,500
Houston Oil Terminal Co.	Houston	1930–1931	3,000
American Petroleum Co.	Norsworthy	1931–1941	10,000
Houston Oil Co. of Texas	Viola	1931–1945	1,000
Stone Oil Co.	Texas City	1931–1948	7,500
Petroleum Conversion Co.	Texas City	1930–1932	500
Conroe Rfg. Co.	Conroe	1933–1935	500
Maritime Oil Co.	Galena Park	1934–1949	5,000
Phoenix Rfg. Co.	Houston	1934–1947	2,500
Bennett Rfg. Co.	Houston	1936–1939	2,500
Washington Rfg. Co.	Conroe	1936–1936	500
MacCook Rfg. Co.	Houston	1937–1937	300
Shelly Rfg. Co.	Bay City	1937–1945	1,200
Alpine Rfg. Co.	Houston	1938–1938	2,500
Green's Bayou Rfg. Co.	Houston	1939–1939	1,000
Hamman Exploration Co.	Bay City	1941–1951	3,000
Eddy Refinery	Houston	1947–1961	2,500
American Republic Corp.	Houston	1947–1948	6,500
Sid Richardson	Texas City	1947–1956	20,000
J.S. Abercrombie Co.	Houston	1947–1947	18,000
Petrol Rfg. Inc.	Texas City	1948–1951	50,000
W. C. McBride, Inc.	Bishop	1951–1956	2,000
Lion Oil Co. (Monsanto)	Texas City	1961–1961	5,000
Southport Pet. Co.	Texas City	1937–1945	20,000
Oil & Chemical Products	Galena Park	1950–1951	2,500

SMALL

Beaumont-Port Arthur-Lake Charles Area:

Refining Company	*Refinery Location*	*Years in Operation*	*Greatest Capacity on Gulf Coast*
United Oil Co.	West Lake, La.	1919–1919	1,000
Southwestern Oil & Rfg.	Lake Charles, La.	1938–1939	5,000
Evangeline	Jennings, La.	1940–1961	2,200
Calcasieu Oil Rfg. Co.	West Lake, La.	1920–1924	700
Hutex Oil & Rfg.	Hardin	1941–1949	1,000
Texas Gas Corp.	Winnie	1959–1961	7,500
Mogul Producing & Rfg.	Orange	1921–1921	3,000
Seaboard Oil & Rfg.	Orange	n/a–1923	500
Southwestern Rfg. Co.	Sour Lake	1924–1927	1,000

Louisiana (Baton Rouge-New Orleans Area):

Refining Company	*Refinery Location*	*Years in Operation*	*Greatest Capacity on Gulf Coast*
Southern	Plaquemine	n/a–1919	500
Pelican Oil & Rfg.	Chalmette	n/a–1922	1,200
Liberty	New Orleans	n/a–1929	4,000
Delart Rfg. Co.	Anse Le Butte	1921–1924	165
Tar Island Producing & Rfg. Co.	Tar Island	1919–1920	60
U.S. Rfg. Co.	Southport	1932–1933	4,000
Meraux Pet. Storage	Meraux	1947–1948	3,000
Chalmette Oil & Rfg.	Chalmette	1922–1949	14,000
General Oil Corp.	Meraux	1940–1945	4,000
Bradley Oil & Rfg.	Eunice	1940–1945	350
Canal Rfg. Co.	Church Pt.	1952–1961	1,500
Southern States Rfg. Co.	Eola	1947–1949	1,000
Breaux Bridge Oil Rfg. Co.	Anse Le Butte	1948–1959	1,000
Gilcrease Oil Co.	Meraux	1949–1951	4,200
Ingram Products Co.	Meraux	1953–1961	22,000

Corpus Christi Area:

Refining Company	*Refinery Location*	*Years in Operation*	*Greatest Capacity on Gulf Coast*
Coastal Refineries	Port Isabel	1936–1949	6,500
Refugio Refinery	Refugio	1936–1947	1,500
Wood-Moore Corp.	Port Lavaca	1936–1939	1,500
Corpus Christi Rfg. Co.	Corpus Christi	1936–1938	12,000
Petrol Coastal Rfg.	Corpus Christi	1937–1937	6,000
Amsco Rfg. Co.	Corpus Christi	1937–1945	11,500
Davis Refinery	Corpus Christi	1937–1938	1,000
Terminal Rfg. Co.	Corpus Christi	1937–1942	12,000
Barnsdall Rfg. Co.	Corpus Christi	1938–1939	5,000

SMALL

Corpus Christi Area (cont'd):

Refining Company	*Refinery Location*	*Years in Operation*	*Greatest Capacity on Gulf Coast*
Alto Rfg. Co.	Corpus Christi	1939-1939	250
Bareco Oil Co.	Corpus Christi	1940-1945	15,500
American Mineral Spirit	Corpus Christi	1940-1949	12,000
Amsco Rfg. Co.	Port Lavaca	1941-1945	1,200
Bennett Oil & Rfg.	Refugio	1948-1949	2,000
Corpus Christi Rfg. Co.	Corpus Christi	1954-1958	10,500
South Texas Rfg. Co.	Brownsville	1956-1959	7,500
Howell Rfg. Co.	Corpus Christi	1959-1961	5,000
Premier Oil Rfg. Co.	Brownsville	1961-1961	5,000

This listing of the remainder of the regional refineries requires several qualifications. First, I have relied on the Bureau of Mines' yearly listings. Thus, my list, like theirs, might omit refineries that were not in operation a complete year. My listing of "years in operation" is actually a listing of the years in which the individual refineries appeared in the Bureau of Mines Information Circulars. Even if the capacity was shown as "shut down," I included that year as a year in operation, since the refinery had reported to the Bureau of Mines.

Those refineries in operation before 1918 received a "n/a" for their first year in operation. Those that disappeared during World War II are given a "1945" as their last year of operation.

The Small refiners are a mixed group. Many did not have more than a single year in which they reported an operating capacity. Some were able to survive for numerous years. Such healthy plants were generally purchased by larger companies.

In differentiating between Large and Small in borderline cases, I relied on the historical context. Thus 10,000 b/d capacity refineries built in the 1920s to handle Mexican crude were assigned to the Large category because they were relatively large for their time and they would probably have grown much larger had their source of crude remained. Conversely, several 15,000–20,000 b/d refineries in the post-World War II period were assigned to the Small category, since the scale of refining had grown and the specific refineries did not experience sustained growth and keep up with the growth of existing plants.

Bibliography

MANUSCRIPT COLLECTION

Austin, Texas. Sid Richardson Building. University of Texas. Eugene C. Barker Texas History Center. Archives Collection. Louis Wiltz Kemp Papers.

Austin, Texas, Sid Richardson Building. University of Texas. Eugene C. Barker Texas History Center. Archives Collection. Oral History of the Texas Oil Industry.

Austin, Texas, Sid Richardson Building. University of Texas. Eugene C. Barker Texas History Center. Texas Collection.

Austin, Texas. Texas State Archives Building. Library Archives Division.

Austin, Texas. Texas State Archives Building. Texas Documents Section.

Beaumont, Texas. Beaumont Public Library. Tyrrell Historical Collection.

Beaumont, Texas. Mobil Oil Refinery. Beaumont Works Library.

Deer Park, Texas. San Jacinto Monument. San Jacinto Museum of History Association. Papers of George M. Craig.

Houston, Texas. Fondren Library, Rice University. James Lockhart Autry papers.

Houston, Texas. Houston Public Library, Historical Collection.

Houston, Texas. Houston Public Library. The Charles A. Warner Collection.

Houston, Texas. Houston Public Library. Joseph Stephen Cullinan papers.

Houston, Texas. Milam Building. Houston Chamber of Commerce. Library.

New York, New York. The Mobil Oil Corporation. The Mobil Archives.

Port Arthur, Texas. Gales Memorial Library. Historical Collection.

Port Arthur, Texas. Port Arthur Chamber of Commerce. Historical Files.

Suitland, Maryland. National Archives Record Center. Record Group 280. Records of the Federal Mediation and Conciliation Service and of the National Labor Relations Board.

Washington, D.C. National Archives. Record Group 70. Bureau of Mines Records.

Washington, D.C. National Archives. Record Group 122. Federal Trade Commission Records.

Washington, D.C. National Archives. Record Group 312. General Correspondence of the Bureau of Foreign and Domestic Commerce.

Washington, D.C. 2101 L N.W. American Petroleum Institute. American Petroleum Institute Library.

White Plains, New York. The Texas Company. The Texas Company Archives.
Wilmington, Delaware. Eleutherian Mills Historical Library. Sun Oil papers.

GOVERNMENT DOCUMENTS

Federal Trade Commission. *The Advance in Price of Petroleum Products. Washington, D.C.: Government Printing Office, 1920.*

———. *The International Petroleum Cartel.* Washington, D.C.: Government Printing Office, 1952.

———. *Prices, Profits, and Competition.* Washington, D.C.: Government Printing Office, 1928.

State of Texas. Bureau of Labor Statistics. Biennial Report, 1909-1910, 1911-1912, 1913-1914, 1915-1916.

———. Office of the Comptroller. Annual Report of the Comptroller of Public Accounts of the State of Texas, 1900.

———. Secretary of State. Biennial Report of the Secretary of State, 1895-1896, 1897-1898, 1899-1900, 1901-1902.

Temporary National Economic Committee. *Investigation of Concentration of Economic Power, Hearings on Public Resolution No. 113,* Parts 14, 14-A, 15, 15-A, 16, 17, 17-A, 75th Congress, 1940.

United States Bureau of the Census. Statistics of Manufactures 1880, II. *Census of Manufactures,* 1890, VI, Part I; 1900, VII, Part I; 1910, IX; 1914, I; 1920 IX; 1930, III; 1940, III; 1947, III; 1954, III; 1963, III; 1972, IV, V.

———. *Census of Population,* 1870, I; 1880, II, Part I; 1890, II, Part I; 1900, I; 1910, III; 1920, III; 1930, III, Parts 1 and 2; 1940, II, Parts 3 and 6; 1950, II, Parts 18 and 43; 1960, I, Parts 20 and 45; 1970, I, Parts 20 and 45.

United States Bureau of Corporations. *Report of the Commissioner of Corporations on the Petroleum Industry: Part I, Position of the Standard Oil Company in the Petroleum Industry; Part II, Prices and Profits.* Washington, D.C. Government Printing Office, 1907.

United States Department of Commerce. *United States Petroleum Refining—War and Post War.* Washington, D.C.: Government Printing Office, 1947.

United States Department of the Interior, Bureau of Mines. *Pollution of the Coast Waters of the United States.* Washington, D.C.: Government Printing Office, 1923.

———. "Petroleum Refineries in the United States, January 1, 1926," 1926. Also, reports of the same title for January 1, 1927; January 1, 1928, Information Circular 6065, April 1928; January 1, 1929, Information Circular 6116, April 1929; January 1, 1930, Information Circular 6292, April 1930; January 1, 1931, Information Circular 6485, May 1931; January 1, 1932, Information Circular 6641, July 1932.

———. "Petroleum Refineries, Including Cracking Plants, in the United States," January 1, 1933, Information Circular 6728, June 1933. Also, reports of the same title for January 1, 1934, Information Circular 6807, August 1934; January 1, 1935, Information Circular 6850, September 1935; January 1, 1936, Information Circular 6906, September 1936; January 1, 1937, Information Circular 6977, December 1937; January 1, 1938, Information Circular 7034, August 1938; January 1, 1939, Information Circular 7091, October 1939; January 1, 1940, Information Circular 7124, June 1940; January 1, 1941, Information Circular 7161, March 1941; January 1, 1947, Information Circular 7455, March 1948; January 1, 1948, Information Circular 7483; August 1948; January 1, 1949, Information Circular 7537, September 1949; January

1, 1950, Information Circular 7578, August 1950; January 1, 1951, Information Circular 7613, July 1951; January 1, 1956, Information Circular 7761, August 1956; January 1, 1961, Information Circular 8062, July 1961.
______. "Petroleum Refineries in the United States and Puerto Rico," January 1, 1966, Mineral Industry Surveys, July 8, 1966. Also, for January 1, 1971, September 14, 1971, and January 1, 1976, July 14, 1976.

BOOKS AND ESSAYS

Published Sources

Adams, Graham, Jr. *Age of Industrial Violence, 1910-1915.* New York: Columbia University Press, 1966.
Allen, Ruth. *Chapters in the History of Organized Labor in Texas.* Austin: University of Texas Press, 1941.
______. *East Texas Lumber Workers; An Economic and Social Picture, 1870-1950.* Austin: University of Texas Press, 1961.
Allin, Benjamin. *Reaching for the Sea.* Boston: Meador Publishing Company, 1956.
Alperin, Lynn M. *Custodians of the Coast: History of the United States Army Engineers at Galveston.* Galveston, Texas: United States Army Corps of Engineers, 1977.
American Guide Series. *Port Arthur.* Houston: Anson Jones Press, 1940.
American Petroleum Institute. *Manual of Disposal of Refinery Wastes, Section I—Waste Water Containing Oil.* New York: A. P. I., 1941.
______. *Manual of Disposal of Refinery Wastes, Section II—Waste Gases and Vapors.* New York: A. P. I., 1931.
______. *Manual of Disposal of Refinery Wastes, Section III-Waste Waters Containing Solutes.* New York: A.P.I., 1935.
______. *Petroleum Facts and Figures,* 1971 Edition. Baltimore: Port City Press, 1971.
______. "Report Covering Survey of Oil Conditions in the United States." New York: American Petroleum Institute, 1927.
Averitt, Robert. *The Dual Economy: The Dynamics of American Industry Structure.* New York: W. W. Norton & Company, Inc., 1968.
Batchelor, Bronson. *Stream Pollution: A Study of Proposed Federal Legislation and Its Effect on the Oil Industry* (prepared for the A. P. I.). New York: American Petroleum Institute, 1937.
Baughman, James. "New Directions in American Economic and Business History." In *American History: Retrospect and Prospect,* edited by Gerald Grob and George Billias. New York: The Free Press, 1971, pp. 271-314.
Bernstein, Irving. *The Lean Years, A History of the American Worker 1920-1933.* Boston: Houghton Mifflin, 1960.
______. *Turbulent Years, A History of the American Worker 1933-1941.* Boston: Houghton Mifflin, 1969.
Berry, Brian J. L. *Growth Centers in the American Urban System.* 2 vols. Cambridge: Ballinger Publishing Company, 1973.
Berry, Brian J. L., and Horton, Frank. *Geographic Perspectives on Urban Systems.* Englewood Cliffs, N.J.: Prentice-Hall, Inc., 1970.

Bragaw, Louis K. et al. *The Challenge of Deepwater Terminals.* Lexington, Mass.: Lexington Books, 1975.

Brody, David. *Steelworkers in America: The Non-Union Era.* Cambridge: Harvard University Press, 1960.

Bryant, Keith. *Arthur E. Stilwell: Promoter With a Hunch.* Nashville: Vanderbilt University Press, 1971.

———. "Arthur Stilwell and the Founding of Port Arthur, A Case of Entrepreneurial Error." *Southwestern Historical Quarterly* LXXV (July 1971), pp. 19–40.

Bureau of Business Research. *Directory of Texas Manufactures, 1947.* Austin: Bureau of Business Research, College of Business Administration, University of Texas at Austin, 1947.

Chandler, Alfred D., Jr. *Strategy and Structure: Chapters in the History of the Industrial Enterprise.* Cambridge: The M.I.T. Press, 1962.

Chandler, Alfred D., Jr. and Galambos, Louis. "The Development of Large-Scale Economic Organizations in Modern America." *Journal of Economic History* XXX (March 1970), pp. 201–217.

Clark, James A., and Halbouty, Michel T. *Spindletop.* New York: Random House, 1952.

Clark, Joseph L. *The Texas Gulf Coast: Its History and Development.* 4 vols. New York: Lewis Historical Publishing Company, Inc., 1955.

Copp, E. Anthony. *Regulating Competition in Oil: Government Intervention in the U.S. Refining Industry,* 1948–1975. College Station, Texas: Texas A&M University Press, 1976.

Davies, J. Clarence, III, and Davies, Barbara. *The Politics of Pollution.* New York: Pegasus, 1975.

Degler, Carl. *The Age of Economic Revolution: 1876–1900.* Glenview, Ill.: Scott Foresman, 1967.

Duecker, Werner W., and West, James R. (eds.). *The Manufactur of Sulfuric Acid.* New York: Reinhold, 1959.

Duncan, Otis; Scott, W. Richard; Lieberman, Stanley; Duncan, Beverly; and Winsborough, Hal. *Metropoi Metropolis and Region.* Baltimore: The Johns Hopkins Press, 1960.

Engler, Robert. *The Politics of Oil, Private Power and Democratic Directions.* New York: The Macmillan Company, 1961.

Enos, John. *Petroleum Progress and Profits: A History of Process Innovation.* Cambridge: M.I.T. Press, 1962.

Fanning, Leonard. *Foreign Oil and the Free World.* New York: McGraw-Hill Book Company, Inc., 1954.

Fishlow, Albert. *American Railroads and the Transformation of the Antebellum Economy.* Cambridge: Harvard University Press, 1965.

Fogel, Robert. *Railroads and American Economic Growth: Essays in Economic History.* Baltimore: The Johns Hopkins Press, 1964.

Forbes, Gerald. *Flush Production, The Epic of Oil in the Gulf-Southwest.* Norman, Oklahoma: University of Oklahoma Press, 1942.

Fornell, Earl. *The Galveston Era: The Texas Crescent on the Eve of Secession.* Austin: University of Texas Press, 1961.

Foscue, Edwin J. "Industrialization of the Texas Gulf Coast Region." *Southwestern Social Science Quarterly* XXXI (June 1950), pp. 1–18.

Frey, John, and Ide, H. Chandler (eds.). *A History of the Petroleum Administration for War, 1941–1945.* Washington: Government Printing Office, 1946.

Fries, Robert F. *Empire in Pine: The Story of Lumbering in Wisconsin, 1830–1900.* Madison: State Historical Society of Wisconsin, 1951.

Fussell, H. R. *A History of Gulf States Utilities Company, 1912-1947.* Houston: Gulf Coast Historical Association, 1967.

Galambos, Louis, with the assistance of Barbara B. Spence. *The Public Image of Big Business in America, 1880-1940, A Quantitative Study in Social Change.* Baltimore: The Johns Hopkins Press, 1975.

Galenson, Walter. *The C.I.O. Challenge to the A.F.L.: A History of the American Labor Movement, 1935-1941.* Cambridge: Harvard University Press, 1960.

Gibb, George S., and Knowlton, Evelyn H. *History of Standard Oil Company (New Jersey): The Resurgent Years, 1911-1927.* New York: Harper & Brothers, 1956.

Gilbert, Chester G., and Pogue, Joseph E. *America's Power Resources.* New York: The Century Co., 1921.

Goldsmith, Raymond (ed.). *Input-Output Analysis: An Appraisal.* Princeton: Princeton University Press, 1955.

Grodinsky, Julius. *Transcontinental Railroad Strategy, 1869-1893: A Study of Businessmen.* Philadelphia: University of Pennsylvania Press, 1962.

Grubb, Herbert. *The Structure of the Texas Economy.* 2 vols. Austin: Office of the Governor, Office of Information Services, 1973.

Gurnham, C. Fred (ed.). *Industrial Wastewater Control: A Textbook and Reference Work.* New York: Academic Press, 1965.

Hamilton, Daniel C. *Competition in Oil: The Gulf Coast Refinery Market, 1925-1950.* Cambridge: Harvard University Press, 1958.

Hansen, Niles (ed.). *Growth Centers in Regional Economic Development.* New York: Free Press, 1972.

Hardwicke, Robert. *Legal History of Gas and Oil.* Chicago, 1938.

Hardy, Dermot, and Roberts, Ingham (eds.). *Historical Review of South-East Texas.* 2 vols. Chicago: Lewis Publishing Co., 1910.

Hart, W. B. "Disposal of Refinery Waste Waters." *Industrial and Engineering Chemistry* XXVI (September 1934), pp. 965-967.

———. *Industrial Waste Disposal for Petroleum Refineries and Allied Plants.* Cleveland: Petroleum Processing, 1947.

Hart, W. B., and Weston, B. F. "The Water Pollution Abatement Problems of the Petroleum Industry." *Water Works and Sewerage* 88 (May 1941), pp. 208-217.

Hawley, Ellis W. "Herbert Hoover, the Commerce Secretariat, and the Vision of an 'Associative State,' 1921-1928." *Journal of American History* LXI (June 1974), pp. 116-140.

———. *The New Deal and the Problem of Monopoly, a Study in Economic Ambivalence.* Princeton: Princeton University Press, 1966.

Haynes, William. *The Stone That Burns: The Story of the American Sulphur Industry.* New York: D. Van Nostrand, 1942.

Hays, Samuel. *Conservation and the Gospel of Efficiency: The Progressive Conservation Movement, 1890-1920.* Cambridge: Harvard University Press, 1959.

Hazelton, Jared E. *The Economics of the Sulphur Industry.* Baltimore: The Johns Hopkins Press, 1970.

Henriques, Robert. *Marcus Samuel.* London: Barrie and Rockliff, 1960.

Hidy, Ralph W. "Business History: Present Status and Future Needs." *Business History Review* XLIV (Winter 1970), pp. 483-497.

Hidy, Ralph W., and Hidy, Muriel E. *History of Standard Oil Company (New Jersey): Pioneering in Big Business, 1882-1911.* New York: Harper & Brothers, 1955.

Hill, R. T. "The Beaumont Oil Field, with Notes on Other Fields of the Texas Region." *Journal of the Franklin Institute,* August-October 1902.

Hirschman, Albert O. *The Strategy of Economic Development.* New Haven: Yale University Press, 1958.

Hofstadter, Richard. "What Happened to the Antitrust Movement? Notes on the Evolution of an American Creed." edited by Earl Cheit, In *The Business Establishment,* pp. 113-113-151. New York: Wiley, 1964.

Isard, Walter; Schooler, E. W.; and Vietorisz, T. *Industrial Complex Analysis and Regional Development: A Case Study of Refinery-Petrochemical-Synthetic Fiber Complexes in Puerto Rico.* New York: John Wiley & Sons, 1959.

Israel, Jerry (ed.). *Building the Organizational Society.* New York: Free Press, 1972.

James, H. G. *Refining Industry of the United States.* Oil City, Pennsylvania: The Derrick Publishing Company, 1916.

James, Marquis, *The Texaco Story: The First Fifty Years, 1902-1952.* New York: The Texas Company, 1953.

Johnson, Arthur. *The Development of American Petroleum Pipe Lines; A Study in Private Enterprise and Public Policy, 1862-1906.* Ithaca: Cornell University Press, 1956.

______. "The Early Texas Oil Industry: Pipelines and the Birth of an Integrated Oil Industry. 1901-1911." *Journal of Southern History* XXXII (November 1966), pp. 516-528.

Kerlin, Gregg, and Rabovsky, Daniel. *Cracking Down: Oil Refining and Pollution Control.* New York: Council on Economic Priorities, 1975.

Kilman, Edward, and Wright, Theon. *High Roy Cullen: A Story of American Opportunity.* New York: Prentice-Hall, Inc., 1954.

King, John O. *The Early History of the Houston Oil Company of Texas, 1901-1908.* Houston: Texas Gulf Coast Historical Association, 1959.

______. *Joseph Stephen Cullinan: A Study of Leadership in the Texas Petroleum Industry, 1897-1937.* Nashville: Vanderbilt University Press, 1970.

Kirksey, C. D. An Interindustry Study of the Sabine-Neches Area of Texas. Austin: Bureau of Business Research, University of Texas at Austin, 1959.

Kuklinski, Antoni (ed.). *Growth Poles and Growth Centres in Regional Planning.* The Hauge: Mouton & Company, 1972.

Landy, Marc Karnis. *The Politics of Environmental Reform: Controlling Kentucky Strip Mining.* Washington, D.C.: Resources for the Future, 1976.

Larson, Agnes. *History of the White Pine Industry in Minnesota.* Minneapolis: University of Minnesota Press, 1949.

Larson, Henrietta; Knowlton, Evelyn; and Popple, Charles. *History of Standard Oil Company (New Jersey): New Horizons, 1927-1950.* New York: Harper & Row, 1971.

Larson, Henrietta, and Porter, Kenneth W. *History of Humble Oil and Refining Company, A Study in Industrial Growth.* New York: Harper & Brothers, 1959.

Leeston, Alfred; Crichton, John; and Jacobs, John. *The Dynamic Natural Gas Industry.* Norman, Oklahoma: University of Oklahoma Press, 1963.

Lightfoot, Jewel. *Anti-Trust Laws of the State of Texas.* Austin: Austin Printing Company, 1907.

Lively, Robert A. "The American System: A Review Article." *Business History Review* XXIX (March 1955), pp. 81-96.

Loos, John L. *Oil On Stream! A History of the Interstate Oil Pipe Line Company, 1909-1959.* Baton Rouge: Louisiana State University Press, 1959.

McComb, David G. *Houston, The Bayou City.* Austin: University of Texas Press, 1969.

McCraw, Thomas K. "Regulation in America: A Review Article." *Business History Review* XLIX (Summer 1975), pp. 159-183.

McDonald, Stephen. *Petroleum Conservation in the United States: An Economic Analysis.* Baltimore: The Johns Hopkins Press, 1971.
McKay, Seth. *Seven Decades of the Texas Constitution of 1876.* Lubbock, Texas: Texas Tech Press, 1942.
______. *W. Lee O'Daniel and Texas Politics, 1938–1942.* Lubbock, Texas: Texas Tech Press, 1944.
McKay, Seth, and Faulk, Odie. *Texas After Spindletop.* Austin: SteckVaughn Co., 1965.
McKee, David; Dean Robert; and Leahy, William (eds.). *Regional Economics: Theory and Practices.* New York: The Free Press, 1970.
McLean, John, and Haigh, Robert. *The Growth of Integrated Oil Companies.* Boston: Division of Research, Graduate School of Business Administration, Harvard University, 1954.
Marcosson, Issac. *The Black Golconda.* New York: Harper & Brothers, 1923.
Marshall, F. Ray. "Independent Unions in the Gulf Coast Petroleum Refining Industry—The Esso Experience." *Labor Law Journal* 12 (September 1961), pp. 823–840.
Mellon, William L. *Judge Mellon's Sons.* Pittsburgh: Privately printed, 1948.
Meyer, John R. "Regional Economics: A Survey." *American Economic Review* 53 (March 1963), pp. 19–54.
Meyer, Lorenzo. *Mexico and the United States in the Oil Controversy, 1917–1942.* Translated by Muriel Vasconcellos. Austin: University of Texas Press, 1977.
Mobil Oil Company Publications Staff. *History of the Refining Department of the Magnolia Petroleum Company.* Beaumont, Texas: Mobil Oil Company, n.d.
Muir, Andrew Forest. "Railroads Come to Houston, 1857–1861." *Southwestern Historical Quarterly* LXIV (July 1960), pp. 42–63.
Nash, Gerald D. *United States Oil Policy, 1890–1964: Business and Government in Twentieth Century America.* Pittsburgh: University of Pittsburgh Press, 1968.
Neuner, Edward. *The Natural Gas Industry: Monopoly and Competition in Field Markets.* Norman, Oklahoma: University of Oklahoma Press, 1960.
Nixon, Edgar B. (ed.). *Franklin D. Roosevelt and Conservation, 1911–1945.* 2 vols. Hyde Park, New York: General Services Administration, National Archives and Record Service, Franklin D. Roosevelt Library, 1957.
Nordhauser, Norman. "Origins of Federal Oil Regulations in the 1920s." *Business History Review* XLVII (Spring 1973), pp. 53–71.
Norvell, G. Todd, and Bell, Alexander. "Air Pollution Control in Texas." In *Legal Control of the Environment,* pp. 337–376. New York: Practicing Law Institute, 1970.
O'Connor, Harvey. *History of the Oil Workers International Union (C.I.O.).* Denver: Oil Workers International Union (CIO), 1950.
Odell, Peter. *An Economic Geography of Oil.* London: Bell, 1963.
Oil Investors' Journal. *Dope Book (1907).* Tulsa: Oil Investors' Journal, 1907.
Oliver, Earl D.; Muller, Robert G.; and Ferguson, Alan F. *The Economic Impact of Environmental Regulations on the Petroleum Industry.* Menlo Park, California: Stanford Research Institute, January 1974.
Olmsted, Frederick Law. *A Journey Through Texas, Or, A Saddle Trip on the Southwestern Frontier.* New York: Dix, Edwards & Co., 1857.
Ozanne, Robert. *A Century of Labor-Management Relations at McCormick and International Harvester.* Madison: University of Wisconsin Press, 1967.
______. *Wages in Practice and Theory: McCormick and International Harvester, 1860–1960.* Madison: Universty of Wisconsin Press, 1968.
Parsons, James J. "Recent Industrial Development in the Gulf South." *Geographical Review* XL (January 1950), pp. 67–83.

Pascal, Anthony H. (ed.). *Racial Discrimination in Economic Life.* Lexington, Mass.: Lexington Books, 1972.

Pelling, Henry. *American Labor.* Chicago: University of Chicago Press, 1960.

Perlman, Selig. *A History of Trade Unionism in the United States.* New York: A. M. Kelly, 1950 [1922].

Perloff, Harvey; Dunn, Edgar, Jr.; Lampard, Eric; Muth, Richard. *Regions, Resources, and Economic Growth.* Baltimore: The Johns Hopkins Press, 1960.

Pittsburgh Regional Planning Association. *Economic Study of the Pittsburgh Region.* 3 vols. Pittsburgh: University of Pittsburgh Press, 1963.

Potter, David. *People of Plenty: Economic Abundance and the American Character.* Chicago: The University of Chicago Press, 1954.

Potts, Charles Shirley. *Railroad Transportation in Texas.* Austin: Bulletin of the University of Texas, 1909.

Pratt, Joseph. "Regional Development in the Contest of National Economic Growth." In *Regional Economic History: The Mid-Atlantic Area since 1700.* Edited by Glenn Porter. Greenville, Wilmington, Delaware: Eleutherian Mills-Hagley Foundation, 1976, pp. 25–40.

Rayback, Joseph. *A History of American Labor.* New York: Macmillan Co., 1959.

Reed, S. G. *A History of Texas Railroads.* Houston: St. Clair Publishing Company, 1941.

Rezler, Julius. "Labor Organization at DuPont: A Study in Independent Local Unionism." *Labor History* IV (Spring 1963), pp. 178–198.

Richardson, Harry. *Regional Growth Theory.* New York: John Wiley & Sons, 1973.

Rister, Carl Coke. *Oil! Titan of the Southwest.* Norman, Oklahoma: University of Oklahoma Press, 1949.

Rundell, Walter, Jr. *Early Texas Oil: A Photographic History, 1866–1936.* College Station, Texas: Texas A&M University Press, 1977.

Schurr, Sam, and Netschert, Bruce. *Energy in the American Economy, 1850–1975: An Economic Study of Its History and Prospects.* Baltimore: The Johns Hopkins Press, 1960.

Seager, Henry, and Gulick, Charles, Jr. *Trust and Corporation Problems.* New York: Harper & Brothers Publishers, 1929.

Sibley, Marilyn McAdams. *The Port of Houston: A History.* Austin: University of Texas Press, 1968.

Smith, Frank E. *The Politics of Conservation.* New York: Harper & Row, 1966.

Smith, Robert Freeman. *The United States and Revolutionary Nationalism in Mexico, 1916–1932.* Chicago: University of Chicago Press, 1972.

Southwestern Legal Foundation. *Economics of the Natural Gas Industry.* New York: Matthew Bender and Company, 1962.

Spratt, John. *The Road to Spindletop: Economic Change in Texas, 1875–1901.* Dallas: S.M.U. Press, 1955.

Startzell, R. H. "Texas Steel." *Texas Journal of Science* II (December 20, 1950), pp. 451–461.

Stewart, W. L., Jr. "Laws, Morals, and Manners." *Proceedings of the 16th Mid-Year Meeting, Division of Refining, A.P.I.* Tulsa, Oklahoma, April 30–May 3, 1951, pp. 11–13.

Stigler, George J. "The Division of Labor is Limited by the Extent of the Market." *Journal of Political Economy* LIX (June 1951), pp. 185–193.

Stockton, James R.; Henshaw, R. C.; and Graves, R. W. *Economics of Natural Gas in Texas.* Austin: Bureau of Business Research, University of Texas at Austin, 1952.

Street, James H. *The New Revolution in the Cotton Economy: Mechanization and Its Consequences.* Chapel Hill: University of North Carolina Press, 1957.

Taft, Philip. *The American Federation of Labor in the Time of Gompers.* New York: Harper, 1957.

———. *The American Federation of Labor from the Death of Gompers to the Merger.* New York: Harper, 1959.

Tanzer, Michael. *The Political Economy of International Oil and the Underdeveloped Countries.* Boston: Beacon Press, 1969.

Thompson, Craig. *Since Spindletop: A Human Story of Gulf's First Half-Century.* Pittsburgh: Gulf Oil Corporation, 1951.

Thompson, Wilbur. *A Preface to Urban Economics.* Baltimore: The Johns Hopkins Press, 1965.

Tindall, George B. *The Emergence of the New South, 1913-1945.* Baton Rouge: Louisiana State University Press, 1967.

Ulman, Lloyd. *The Rise of the National Trade Union.* Cambridge: Harvard University Press, 1955.

Warner, Charles A. *Texas Oil and Gas Since 1543.* Houston: Gulf Coast Publishing Company, 1939.

Westcott, James H. *Oil, Its Conservation and Waste.* New York: Beacon Publishing Company, 1928.

Wheeler, Kenneth. *To Wear a City's Crown: The Beginnings of Urban Growth in Texas, 1836-1865.* Cambridge: Harvard University Press, 1968.

Whitehorn, Norman C. *Economic Analysis of the Petrochemical Industry in Texas.* College Station, Texas: Industrial Economics Research Division, Texas Engineering Experiement Station, Texas A&M University, 1973.

Wiebe, Robert. *The Search for Order, 1877-1920.* New York: Hill and Wang, 1967.

Williams, T. Harry. *Huey Long.* New York: Alfred A Knopf, Inc., 1969.

Williamson, Harold, and Daum, Arnold R. *The American Petroleum Industry: The Age of Illumination, 1859-1899.* Evanston: Northwestern University Press, 1959.

Williamson, Harold; Andreano, Ralph; Daum, Arnold; and Klose, Gilbert. *The American Petroleum Industry: The Age of Energy, 1899-1959.* Evanston: Northwestern University Press, 1963.

Woodward, C. Vann. *Origins of the New South, 1877-1913.* Baton Rouge: Louisiana State University Press, 1951.

Youngberg, J. C. *Natural Gas: America's Fastest Growing Industry.* San Francisco: Schwabacher-Frey, 1930.

Unpublished Sources

Andreano, Ralph. "The Emergence of New Competition in the American Petroleum Industry Before 1911." Ph.D. dissertation, Northwestern University, 1960.

Dixon, Donna. "A History of O.C.A.W. Local 4-228. Port Neches, Texas." M.A. thesis. Lamar College, 1970.

Easton, Hamilton. "The History of the Texas Lumbering Industry." Ph.D. dissertation, University of Texas at Austin, 1947.

Hunt, Lois. "A History of the Beaumont Municipal Port, 1856-1942." M.A. thesis, University of Texas at Austin, 1943.

Kemp, Louis, W., and Talman, Wilfred B. "Documentary History of the Texas Company." New York: The Texas Company, n.d.

McLaughlin, Marvin. "Reflections on the Philosophy and Practices of Lamar State College of Technology." Ph.D. dissertation, University of Houston, 1955.

Martin, Everette. "A Hisory of the Spindletop Oil Field." M.A. thesis, University of Texas at Austin, 1934.

Mitchell, Alfred B. "The Demand for Texas Refined Petroleum Products: A Study in Econometrics." Ph.D. dissertation. University of Texas at Austin, 1971.

Pratt, Joseph. "The Oil Workers International Union's Organization of the Upper Texas Gulf Coast." Senior honors thesis, Rice University, 1970.

Rochelle, John. "Port Arthur: A History of Its Port to 1963." M.A. thesis, Lamar College, 1969.

Trevey, Marilyn. "The Social and Economic Impact of the Spindletop Oil Boom in Beaumont in 1901." M.A. thesis, Lamar College, 1974.

Vague, Thomas A. "An Economic Survey of the Sabine Area, Texas." M.B.A. thesis, University of Texas at Austin, 1949.

Williams, Edward B., Jr. "Pollution Control: A Houston Ship Channel Issure." M.A. thesis, Texas A&M Univeristy, 1972.

Williams, Edward J. "The Impact of Technology on Employment in the Petroleum Refining Industry in Texas, 1947-1966." Ph.D. dissertation, University of Texas at Austin, 1971.

Newspapers and Trade Journals

Beaumont *Age,* 1901.
Beaumont *Enterprise,* 1899-1970.
Beaumont *Journal,* 1901-1970.
Coal Oil Clarion, 1924.
Houston *Chronicle,* 1950-1960.
Houston *Post,* 1920-1965.
The Lamp, 1951.
The Magpetco, 1921-1930.
Manufactures Record, 1899-1905.
National Oil Reporter, 1901-1902.
Oil Investors' Journal, 1902-1910.
Oil and Gas Journal, 1910-1970.
Oil Weekly, 1930-1950.
Petroleum Refiner, 1921-1925.
Port Arthur *Herald,* 1900-1902.
Port Arthur *News,* 1900-1971.
Texaco *Look Box,* 1921-1929.
Texaco *Star,* 1920-1930.

Index